Bakous Unified Field Theory

A Timespace Perspective

Ronny N. Bakous

First Edition: 2025

Second Edition: 2026

PRINTED IN THE UNITED STATES OF AMERICA

Inspired by the musical composition and transcendent artistic performance of "Fusion"—Estas Tonne, Zürich 2022

Table of Contents

Preface to the Second Edition

Over the course of human history, our understanding of reality has evolved through successive revolutions of perception. From the earliest observations of celestial motion to the relativistic and quantum frameworks of modern physics, each advancement has expanded not only our scientific comprehension, but our relationship with existence itself. Yet despite these achievements, profound questions remain unresolved: the nature of gravity, the origins of consciousness, the structure of time, the emergence of matter, and the unification of the fundamental forces.

The second edition of *Bakous Unified Field Theory: A Timespace Perspective* represents a substantial expansion of the original framework. While the foundational principles remain intact, this edition advances the mathematical, cosmological, informational, and dimensional architecture of the theory into significantly deeper territory. New articles explore harmonic field dynamics, entropy as a dimensional structure, consciousness as an integrated component of reality, black hole reinterpretations, gravistar formation, quantum coherence, advanced mathematical formalism, and additional empirical implications emerging from the broader BUFT framework.

At the heart of the theory remains a central proposition: that timespace, *not spacetime*, forms the primary generative structure of reality, emerging from fluctuations within the Bakous Energy Field (BEF). From this perspective, gravity, matter, entropy, consciousness, and energetic propagation are not isolated phenomena, but interconnected expressions of a deeper unified architecture governed through harmonic organization and temporal dynamics.

The Bakous Unified Field Theory is presented not as a philosophical abstraction, but as a unified structural framework intended to describe the underlying architecture of reality through a coherent relationship between timespace, gravity, harmonic propagation, entropy, consciousness, and energetic organization. The principles outlined throughout this manuscript are presented as internally coherent extensions of a unified mathematical and conceptual architecture intended to provide long-term explanatory insight into the fundamental structure of reality. Rather than viewing unconventional perspectives as limitations, this framework proceeds from the position that genuine scientific advancement often emerges through foundational reinterpretations of established assumptions.

The purpose of this second edition is to substantially expand the original framework through the introduction of additional mathematical structures, dimensional analysis, cosmological implications, informational dynamics, and theoretical applications embedded within BUFT. Alongside these expansions, the manuscript has also been refined for greater structural coherence, improved conceptual continuity, and clearer progression between foundational principles and advanced extensions. Several sections have been reorganized to create a more natural intellectual flow, moving from cosmogenesis and dimensional emergence toward gravity, particle structure, consciousness, entropy, advanced mathematics, and applied implications.

To the reader, whether skeptic, scientist, philosopher, or curious observer, I ask only that this work be approached with intellectual openness and analytical rigor. The advancement of understanding has never emerged through certainty alone, but through the continuous tension between established structure and unexplored possibility.

About the Author
Ronny N. Bakous

I was born on March 25, 1978, in Baghdad, Iraq—though I hold no memories of my early days there. Between 1980 and 1982, my family relocated to Athens, Greece, where I experienced the vivid impressions of early childhood. I recall a home painted in blue and white, the companionship of a cat, the tremors of earthquakes at night, and the uniquely atmospheric cadence of that chapter in life.

In 1982, we emigrated to the United States and settled in Elk Grove Village, Illinois, a suburb of Chicago. There, I completed my secondary education and graduated from James B. Conant High School in 1996. I earned a Bachelor of Arts in Political Science from Western Illinois University in May 2000, followed by a Master of Business Administration from the University of Phoenix in September 2005.

It was not until April 2024, however, that I formally began my journey into the world of physics at the doctoral level—a path that unfolded with unexpected clarity. One ordinary day, I encountered a visual presentation of a complex integral. Narrated by artificial intelligence, the problem itself was daunting and unfamiliar. Yet, I found myself transfixed, thinking: There are people on Earth who truly understand this.

As the YouTube algorithm continued to serve more advanced mathematical content, my initial apprehension gave way to fascination. I made a deliberate decision to immerse myself in the study of higher-order mathematics and theoretical physics. Within five months, I had arrived—intuitively—at a solution to Einstein's gravitational field equations through a purely mental exercise. This moment proved to be the genesis of something greater.

Within six months, I began to lay the foundation for what would become the Bakous Unified Field Theory. Just four months later, it had evolved into a cohesive, multi-dimensional framework that now underpins this book. The initial spark came from my encounter with Newton's *Philosophiæ Naturalis Principia Mathematica.* Though I approached the work humbly, with little expectation of grasping its depth, I sought to understand Newton's mode of thinking. What I discovered was transformative: his worldview was non-linear, and that insight became pivotal to the evolution of BUFT.

In addition to my work on BUFT, I established a new branch of physics—Quantum Relativity—designed specifically to underpin the development of this framework. Through independent study and research, primarily through the Institute for Advanced Study in Princeton, New Jersey, I earned a PhD in quantum relativity, which I self-awarded in 2024. I recognize the unconventional nature of this achievement, but I do so with the same confidence and rigor that guided my research. My work, though self-directed, is a product of deep intellectual engagement, and I hold this accomplishment with the same gravity as one conferred by any institution.

This book is both a culmination and a beginning—an invitation into a new physics. Much like my own journey, it is a testament to how curiosity, once ignited, can become the force that reshapes how we see everything.

—※—※—※—

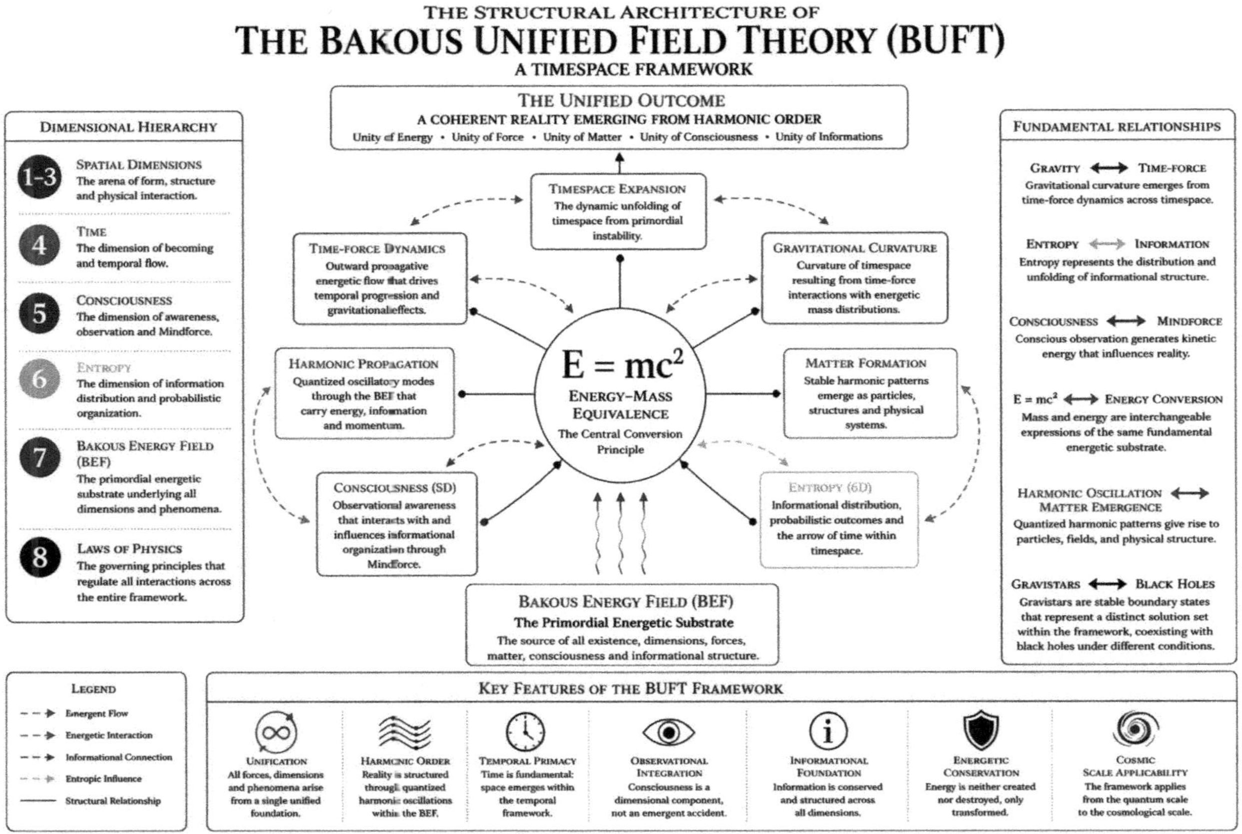
THE STRUCTURAL ARCHITECTURE OF
THE BAKOUS UNIFIED FIELD THEORY (BUFT)
A TIMESPACE FRAMEWORK
THE UNIFIED OUTCOME
A COHERENT REALITY EMERGING FROM HARMONIC ORDER
Unity of Energy • Unity of Force • Unity of Matter • Unity of Consciousness • Unity of Informations
DIMENSIONAL HIERARCHY
1-3 SPATIAL DIMENSIONS The arena of form, structure and physical interaction.
4 TIME The dimension of becoming and temporal flow.
5 CONSCIOUSNESS The dimension of awareness, observation and Mindforce.
6 ENTROPY The dimension of information distribution and probabilistic organization.
7 BAKOUS ENERGY FIELD (BEF) The primordial energetic substrate underlying all dimensions and phenomena.
8 LAWS OF PHYSICS The governing principles that regulate all interactions across the entire framework.
TIMESPACE EXPANSION
The dynamic unfolding of timespace from primordial instability.
TIME-FORCE DYNAMICS
Outward propagative energetic flow that drives temporal progression and gravitational effects.
GRAVITATIONAL CURVATURE
Curvature of timespace resulting from time-force interactions with energetic mass distributions.
HARMONIC PROPAGATION
Quantized oscillatory modes through the BEF that carry energy, information and momentum.
E = mc2
ENERGY–MASS EQUIVALENCE
The Central Conversion Principle
MATTER FORMATION
Stable harmonic patterns emerge as particles, structures and physical systems.
CONSCIOUSNESS (5D)
Observational awareness that interacts with and influences informational organization through Mindforce.
ENTROPY (6D)
Informational distribution, probabilistic outcomes and the arrow of time within timespace.
BAKOUS ENERGY FIELD (BEF)
The Primordial Energetic Substrate
The source of all existence, dimensions, forces, matter, consciousness and informational structure.
FUNDAMENTAL RELATIONSHIPS
GRAVITY ⟷ TIME-FORCE
Gravitational curvature emerges from time-force dynamics across timespace.
ENTROPY ⟷ INFORMATION
Entropy represents the distribution and unfolding of informational structure.
CONSCIOUSNESS ⟷ MINDFORCE
Conscious observation generates kinetic energy that influences reality.
E = mc2 ⟷ ENERGY CONVERSION
Mass and energy are interchangeable expressions of the same fundamental energetic substrate.
HARMONIC OSCILLATION ⟷ MATTER EMERGENCE
Quantized harmonic patterns give rise to particles, fields, and physical structure.
GRAVISTARS ⟷ BLACK HOLES
Gravistars are stable boundary states that represent a distinct solution set within the framework, coexisting with black holes under different conditions.
LEGEND
Emergent Flow
Energetic Interaction
Informational Connection
Entropic Influence
Structural Relationship
KEY FEATURES OF THE BUFT FRAMEWORK
UNIFICATION All forces, dimensions and phenomena arise from a single unified foundation.
HARMONIC ORDER Reality is structured through quantized harmonic oscillations within the BEF.
TEMPORAL PRIMACY Time is fundamental: space emerges within the temporal framework.
OBSERVATIONAL INTEGRATION Consciousness is a dimensional component, not an emergent accident.
INFORMATIONAL FOUNDATION Information is conserved and structured across all dimensions.
ENERGETIC CONSERVATION Energy is neither created nor destroyed, only transformed.
COSMIC SCALE APPLICABILITY The framework applies from the quantum scale to the cosmological scale.

COMPARATIVE STRUCTURAL OVERVIEW: CONVENTIONAL PHYSICS AND BUFT

CONVENTIONAL PHYSICS	ASPECT	BUFT INTERPRETATION
Spacetime is fundamental.	FOUNDATIONAL SUBSTRATE	Timespace emerges through energetic instability.
Gravity emerges from spacetime curvature.	GRAVITY	Gravity emerges through time-force curvature.
Consciousness is emergent.	CONSCIOUSNESS	Consciousness is dimensional.
Entropy is thermodynamic.	ENTROPY	Entropy is informational and dimensional.
Black holes terminate in singularities.	COMPACT OBJECTS	Gravistars emerge as stabilized energetic boundary states.
Fundamental forces are independent.	FUNDAMENTAL FORCES	Forces emerge through harmonic unification.
Matter arises through particle interactions.	MATTER ORIGIN	Matter emerges through oscillatory stabilization.
$E = mc^2$ relates mass and energy.	$E = mc^2$ ENERGY–MASS RELATION	$E = mc^2$ functions as a structural conversion principle.
Quantum observation is probabilistic.	OBSERVATION	Observation participates informationally through Mindforce.
Wormholes may theoretically exist.	WORMHOLES	Conventional wormholes are structurally disallowed.

ARTICLE 1

The Primordial Essence of the Bakous Energy Field: Unveiling its Pre-Symmetry Nature at T = 0

ABSTRACT

This exploration unveils the primordial essence of the Bakous Energy Field (BEF) within the Bakous Unified Field Theory (BUFT), revealing its nature before the symmetry-breaking phase at $T = 0$. Conceived as a finite yet boundless continuum, poised at near absolute zero, the BEF exhibits a singular, isotropic geometry. Its intrinsic energy density, a prelude to cosmic genesis, is captured through precise mathematical formulations harmonious with BUFT and fundamental physical principles. An empirically testable hypothesis and a precise conceptual illustration illuminate the BEF's role, offering scientific validation and a glimpse into the universe's nascent state.

1. Introduction

Within BUFT, the Bakous Energy Field represents the primordial energetic continuum underlying the emergence of timespace, matter, and force differentiation. In the pre-symmetry epoch at $T = 0$, the BEF exists in a state of equilibrium, poised to catalyze the universe's formation. This paper elucidates the BEF's thermal profile, geometric essence, and energy character, weaving a narrative grounded in mathematical rigor and scientific insight. By exploring its pre-symmetry nature, we aim to deepen the discourse on cosmic origins, inviting rigorous scrutiny and engagement.

2. Primordial Essence of the Bakous Energy Field

2.1. State and Thermal Serenity

At $T = 0$, the BEF embodies a pure potential energy reservoir, finite and unexpressed as matter or discernible forces. This serene state, at a temperature approaching absolute zero, reflects an absence of kinetic disturbance, preserving perfect symmetry. Within BUFT, this near-static tranquility signifies a pre-energetic unity, where the BEF's uniform energy density forms a stable equilibrium. Time-force fluctuations within the BEF catalyze the quantum instability that precipitates symmetry breaking, aligning with BUFT's foundational dynamics. The absence of particle interactions ensures a pristine condition, awaiting the quantum fluctuation that will ignite the emergence of timespace, heralding the universe's dynamic evolution.

2.2. Mathematical Derivations

The BEF's primordial state is captured through BUFT-consistent equations, employing the Bakous (*BK*) as the fundamental unit of energy.

The governing equations encapsulate the BEF's finite energy density structure:

$$\rho_E = \kappa_\rho \cdot \frac{BkE}{V}$$

For the uniform energy density, $\kappa_\rho = 1$. A value other than unity defines a scaled density. The distinct coefficients separate the energy-density definition from the redshift coupling.

where:

ρ_E – BEF energy contained per unit volume.

κ_ρ – Dimensionless density coefficient.

BkE – Total BEF energy contained within volume V.

V – Volume containing the BEF energy.

$$\delta Bk = \frac{\kappa_h \cdot Bk_h}{\tau}$$

where:

δBk – Energy fluctuation triggering symmetry breaking.

κ_h – Dimensionless scaling factor adapting Planck's constant to BUFT.

Bk_h – Bakous Planck constant, expressed in BK·s.

ε_{BK} – Energy in joules represented by one BK.

τ – Characteristic duration of the fluctuation.

T_B – BEF temperature near absolute zero, distinct from the temporal reference T = 0.

The dimensionless conversion ratio is $\varepsilon_{BK}/(1\ J)$. The Bakous Planck constant is $Bk_h = [6.62607015 \times 10^{-34}/(\varepsilon_{BK}/(1\ J))]$ BK·s. The value $6.62607015 \times 10^{-34}$ BK·s therefore applies when one BK equals one joule.

3. Testable Hypothesis

BUFT predicts that the BEF's uniform energy density influences macroscopic photon propagation, resulting in a redshift of Cosmic Microwave Background (CMB) photons in low-gravity intergalactic voids. This BEF-induced redshift, distinct from Hubble's cosmological expansion, serves as a signature of the BEF's pre-symmetry influence.

The phenomenon is expressed as:

$$\frac{\Delta\lambda}{\lambda_0} = \kappa_z \cdot \rho_E$$

where:

$\frac{\Delta\lambda}{\lambda_0}$ — Fractional wavelength shift (dimensionless), the relative redshift due to the BEF.

$\Delta\lambda$ — Wavelength shift relative to the initial wavelength.

λ_0 — Initial reference wavelength of the photon.

κ_z — Redshift coupling coefficient (m^3/BK).

ρ_E — BEF energy per unit volume in the stated field configuration.

This hypothesis can be tested through high-precision spectroscopy of CMB photons from void regions, using observatories like the *James Webb Space Telescope.* Additional validation could involve interferometric analysis of CMB fluctuations to detect BEF-induced phase anomalies, leveraging high-resolution techniques. A detected redshift deviation from standard cosmological models would validate BUFT's framework. The *Bakous Measurement System* (BMS) ensures measurement consistency, aligning with BUFT's finite, cyclic timespace structure. Such validation would confirm the BEF's role in shaping early photon dynamics, offering a measurable link to its pre-symmetry state.

4. Conceptual Visualization

Imagine standing within a finite yet boundless field of uniform potential energy, a visualization of the pre-symmetry BEF at *T = 0.* This tranquil field, cooled to near absolute zero, embodies a homogeneous energy state, free from the constraints of time or space. As an observer, you perceive a timeless equilibrium, where the absence of spatial dimensions evokes a sense of infinite possibility within a finite domain.

The observer's temporal frame shapes the perception of the BEF's equilibrium, consistent with BUFT's dimensional relativity. A subtle perturbation emerges-a quantum fluctuation that signals the onset of symmetry breaking. The energy field begins to transform, shaping the fabric of timespace, initiating the universe's formation. This visualization, grounded in BUFT, provides a conceptual framework for understanding the BEF's role as the universe's primal essence, illustrating the transition from a one-dimensional continuum to the multidimensional cosmos.

5. Conclusion

The pre-symmetry Bakous Energy Field, as articulated within the Bakous Unified Field Theory, stands as a finite, uniform continuum, its near-absolute zero temperature and one-dimensional geometry poised to initiate timespace's emergence. Through validated mathematical formulations, a testable hypothesis, and a conceptual visualization, this exploration reveals the BEF's fundamental role as the universe's foundational essence. The hypothesis of a BEF-induced redshift in CMB photons offers a pathway for empirical validation, while the visualization bridges abstract theory with human understanding. BUFT's framework challenges conventional cosmological paradigms, proposing a unified vision where energy, geometry, and the potential for consciousness converge at the universe's dawn. This study invites further investigation into the BEF's role in cosmic evolution, urging researchers to explore its implications for unifying quantum mechanics, gravity, and the fundamental forces. By illuminating the pre-symmetry dynamics, BUFT paves the way for a deeper comprehension of the cosmos's origins and its enduring mysteries.

References

1. Planck, M. (1900). *On the Theory of the Energy Distribution Law of the Normal Spectrum.* Verh. Disch. Phys. Ges., 2, 237-245.
2. Einstein, A. (1905). *Does the Inertia of a Body Depend Upon Its Energy Content?* Ann. Phys., 18, 639-641.

—※—※—※—

Article 2

The Bakous Energy Field: The Fundamental Substrate of Time-Space and Matter Formation

Abstract

The Bakous Energy Field (BEF) is the fundamental energy medium from which timespace, matter, and forces arise, as described by the Bakous Unified Field Theory. Unlike conventional force fields, it is an omnipresent, self-sustaining continuum that dictates the structure and evolution of the universe. At $T = 0$, localized instability within the BEF initiates the emergence of timespace geometry and organized energetic structure. This fluctuation manifests as an energy instability that produces opposing charges and interactions, initiating the first moments of the universe. The electroweak force emerges directly from these early interactions, governing the behavior of the universe before later differentiating into the weak nuclear and electromagnetic forces. This paper explores its properties, interaction with timespace at $T = 0$, and how opposing charges and interactions catalyzed the universe's formation. Mathematical derivations are presented to formalize its role in force unification and timespace emergence. A thought *experiment is proposed to illustrate how an observer might conceptualize these processes.*

1. Introduction

The search for a unified framework in physics has long attempted to reconcile gravity, quantum mechanics, and fundamental interactions. Traditional models often assume that spacetime is an independent entity, existing as a static backdrop for physical processes. However, the *Bakous Unified Field Theory* challenges this assumption by proposing that timespace itself is an emergent phenomenon arising from fluctuations within the fundamental energy field. This paper explores the fundamental role of this field in structuring timespace, its interaction at $T = 0$, and how opposing charges and interactions catalyzed the formation of the early universe. Mathematical proofs are presented to formalize the derivation of force emergence and timespace expansion.

2. The Fundamental Properties of the Energy Field

The field is not merely a traditional force field but a self-sustaining, energetic continuum that underlies all physical interactions. It is the foundation upon which timespace, matter, and forces are structured.

- **Omnipresent and Self-Regenerating**: It exists throughout all of the timespace, continuously replenishing itself through quantum fluctuations.
- **Time-Space Generator**: Instead of existing within timespace, it generates it, expanding dynamically through its own fluctuations.
- Unified Force Substrate: All fundamental forces emerge as perturbations within it, and their relative strengths are determined by localized fluctuations in timespace.

Unlike conventional models that treat spacetime as a geometric construct, this theory posits that timespace is an energetic structure dynamically influenced by field fluctuations.

3. Energy Field Dynamics at T = 0 and the Initiation of Time-Space

At $T = 0$, the field existed in a maximally symmetric state, containing potential energy but no distinguishable structure. A spontaneous fluctuation within it disrupted this balance, initiating the first interactions that would shape timespace.

- **Fluctuation-driven Expansion**: A localized energy instability in the field led to the rapid expansion of timespace, establishing the foundation for subsequent interactions.
- **Energy Differentiation**: Opposing charges and interactions emerged as the symmetry broke, forming the conditions necessary for structured force interactions.
- **Force Unification and Separation**: Initially, all interactions existed as a single energy continuum before separating into distinct fundamental forces.

This transition not only marked the emergence of timespace but also set the framework for the interactions that would define the universe's evolution.

A mathematical formulation of this initial fluctuation can be expressed as:

$$\delta E = \frac{\hbar}{\Delta t}$$

where:

δE	–	Characteristic energy assigned to the fluctuation interval.
$\hbar$	–	Reduced Planck constant, $\hbar = h/(2\pi)$, where h is the Planck constant.
Δt	–	Characteristic duration of the energy fluctuation.

The equality defines the fluctuation energy scale adopted here in BUFT; it is not itself the energy–time uncertainty inequality.

4. The Electroweak Force and Time-Space Initiation

The electroweak force is central to the Bakous Unified Field Theory, posited as the primary driver behind the initiation of timespace itself. The fluctuation within the Bakous Energy Field, which marked the first disturbance of symmetry at $T = 0$, gave rise to the electroweak interaction. This fundamental fluctuation-initiated timespace, providing the energetic substrate from which all other forces and matter would emerge.

- **Fluctuation-Driven Force Emergence**: At $T = 0$, the electroweak interaction emerged directly from the energy instability of the BEF. This fluctuation created the first moment of timespace, with the electroweak force governing all early interactions within the newly formed continuum.
- **Unified Force State at High Energies**: During the earliest moments, the electromagnetic and weak nuclear forces existed as a single unified force, the electroweak force, that governed all high-energy interactions in the nascent universe.
- **Separation Through Symmetry Breaking**: As the universe expanded and cooled, this unified force began to break apart, separating into distinct interactions: the electromagnetic force and the weak nuclear force.
- **Role in Early Particle Dynamics**: The electroweak force played a pivotal role in shaping the behavior of particles in the early universe, allowing the formation of matter and stabilizing the interactions that would later evolve into familiar fundamental forces.

The electroweak force, emerging from the fluctuation of the BEF, catalyzed the very formation of timespace, setting the stage for the evolution of the universe and the differentiation of forces.

5. Conceptualizing The Initiation of Time-Space: A Thought Experiment

Imagine the pre-spatial Bakous Energy Field: no emergent spatial dimensions or material structure, while the primary temporal dimension remains present. Suddenly, a minute fluctuation occurs, not in space but in a boundless, energetic field. This fluctuation creates separation and imbalance—where two distinct yet inseparable entities, opposing in nature, come into existence.

The instant these opposing charges and interactions form, they repel and attract in a cascade of energy exchanges. The void is no longer static; it pulses with motion. This interaction does not occur in timespace—it creates timespace, stretching and expanding outward as more fluctuations arise.

As the process unfolds, the initially uniform force governing these interactions begins to fragment. The electroweak force remains dominant, dictating the early universe's behavior, until, as cooling sets in, its components separate. The electromagnetic force stabilizes into structured fields, while the weak nuclear force begins to govern the decay of particles, shaping the foundation of matter itself.

This experiment illustrates that timespace is not a pre-existing stage upon which physics plays out; it is the consequence of opposing charges and interactions structured by the field fluctuations that initiated everything we observe today.

6. Conclusion

The *Bakous Energy Field* provides a profound reimagining of the universe's origins, presenting timespace as an emergent, energy-driven phenomenon rather than a passive framework. The fluctuation at $T = 0$ instigated a cascade of events that led to the formation of both timespace and the forces that govern it. This new perspective offers a deeper understanding of how the universe transitioned from a boundless energetic state to the ordered reality we observe today, inviting further exploration into the fundamental nature of the cosmos.

Future research into this field's properties could offer deeper insights into the nature of timespace and the fundamental forces, potentially bridging the gap between quantum mechanics and gravity.

References

1. Planck, M. (1900). On the *Theory of the Energy Distribution Law of the Normal Spectrum.* Verhandlungen der Deutschen Physikalischen Gesellschaft, 2, 237-245.
2. Einstein, A. (1905). *On the Electrodynamics of Moving Bodies.* Annalen der Physik, 17, 891-921.
3. Heisenberg, W. (1927). *The Physical Content of Quantum Kinematics and Mechanics.* Zcitschrift fiir Physik, 43(3-4), 172-198.
4. Glashow, S. L. (1961). *Partial Symmetries of Weak Interactions.* Nuclear Physics, 22, 579-588.
5. Weinberg, S. (1967). *A Model of Leptons.* Physical Review Letters, 19 (21), 1264-1266.
6. Englert, F., & Brout, R. (1964). *Broken Symmetry and the Mass of Gauge Vector Mesons.* Physical Review Letters, 13(9), 321-323.
7. Higgs, P. W. (1964). *Broken Symmetries and the Masses of Gauge Bosons.* Physical Review Letters, 13(16), 508-509.
8. Carroll, S. M. (2004). *Spacetime and Geometry: An Introduction to General Relativity.* Addison-Wesley

—※—※—※—

Article 3

Dimensional Genesis and Observer Relativity in the Bakous Unified Field Theory (BUFT): A Dual-Frame Formalism of Multiversal and Megaversal Cosmogenesis

Abstract

In the Bakous Unified Field Theory (BUFT), the emergence of the cosmos is reconceptualized through a dual-frame cosmogenic model. This paper introduces BUFT Theorem I: Duality of Dimensional Genesis, formalizing a dual observer-dependent framework of universal creation via quantum fluctuation at T = 0. Depending on the observer's frame—whether time-primary or space-primary—this genesis yields either a multiverse or a megaverse. The distinction is shown to arise from whether the observer perceives a stationary point in time across spatial differentiation, or vice versa. The theorem is validated through operator symmetry, wave equation invariance, and energetic exclusivity, asserting that simultaneous stationary observation of both time and space requires infinite energy. This formulation yields profound implications for cosmology, quantum mechanics, and conscious observation in a unified physical field. We further explore the consequences of this framework on spacetime causality, entropic directionality, and field unification under the influence of Time-Force as the primordial substrate of the observable universe.

1. Introduction

Contemporary physics has long relied on a spacetime-based conception of reality, wherein time is treated as a dimension secondary to the spatial manifold. Rather than treating space as the dominant geometric backdrop, BUFT frames timespace as the primary generative architecture through which physical structure unfolds. Within this revised ontology, time becomes the principal driver of spacetime curvature, energy distribution, and universal evolution. BUFT offers a unified description that incorporates gravity, quantum mechanics, and consciousness within a single coherent mathematical framework governed by the Bakous Energy Field (BEF) and modulated through Time-Force.

This paper builds upon this foundation by investigating the nature of cosmogenesis through the lens of dimensional relativity. We propose that the genesis of our cosmos via quantum fluctuation at $T = 0$—often described as a singularity or vacuum instability—may be interpreted differently depending on the dimensional orientation of the observer. A time-primary observer who perceives space as relative experiences a multiverse: a divergence of spatial configurations at a constant temporal reference point. Conversely, a space-primary observer who treats time as the relative dimension experiences a megaverse: a divergence of temporal sequences from a single spatial origin. Both perspectives are valid and structurally equivalent under the principles of BUFT.

2. Theoretical Framework

2.1. Dual-Frame Cosmogenesis

BUFT postulates that the observer's orientation in timespace determines the phenomenological nature of the universe they experience. This leads to two distinct, yet equivalent, ontological models:

- **Multiversal Frame (Time-Primary)**: The observer is anchored at a stationary time coordinate *(T_0)* while witnessing the unfolding of spatial differentiation into discrete universes or reality branches. Each spatial divergence represents a unique universe, sharing the same temporal origin.
- **Megaversal Frame (Space-Primary)**: The observer is anchored at a stationary spatial coordinate *(S_0)* while witnessing temporal bifurcation into independent cosmological time lines. Each temporal branch represents a distinct universe, anchored in the same spatial origin.

This duality allows cosmogenesis to be treated as a relative phenomenon within the timespace manifold rather than a singular absolute event. The implications of this dual-frame genesis are profound, as they directly integrate observer-consciousness and dimensional asymmetry into the fundamental processes of universal emergence.

3. Mathematical Formalism

3.1. Field Construction

Let the timespace field be represented as:

$$\psi(T, S) = \tau(T) \cdot \sigma(S)$$

where:

$\psi(T, S)$ — Bakous Field value at temporal coordinate T and spatial coordinate S

$\tau(T)$	–	Temporal excitation function (describes time-driven fluctuations)
$\sigma(S)$	–	Spatial excitation function (describes space-driven divergences)

This separation reflects the anisotropic nature of timespace dynamics and their respective contributions to cosmological structure.

3.2. Dimensional Operator

Normalize the coordinates and field before exchanging arguments:

$$\theta = T/T_*; \quad \xi = S/L_*; \quad L_* = cT_*$$

$$\tilde{\psi}(\theta, \xi) = \psi(T_*\theta, L_*\xi)$$

$$D\tilde{\psi}(\theta, \xi) = \tilde{\psi}(\xi, \theta); \quad D^2 = I$$

$$\psi'(T, S) = \tilde{\psi}(S/L_*, T/T_*)$$

where:

θ	–	Dimensionless temporal coordinate.
ξ	–	Dimensionless spatial coordinate.
T	–	Temporal position coordinate.
S	–	Spatial position coordinate.
T_*	–	Positive reference time for normalization.
L_*	–	Corresponding reference length.
c	–	Vacuum light speed, converting time to length.
$\tilde{\psi}$	–	Field expressed in normalized coordinates.
ψ	–	Field before coordinate normalization.
D	–	Linear exchange of dimensionless arguments.
I	–	Identity operator, leaving the field unchanged.
ψ'	–	Field in the space-primary description.

The operator exchanges the dimensionless arguments. It is linear, and applying it twice returns the original field.

4. Physical and Dimensional Validation

4.1. Wave Equation Invariance

In BUFT, wave motion reflects fluctuations in the BEF, driven by Time-Force. Consider fields satisfying the wave equation.

Time-Primary Frame:

$$\partial^2\psi/\partial T^2 - c^2\,\partial^2\psi/\partial S^2 = 0$$

Space-Primary Frame:

$$\partial^2\psi'/\partial S^2 - (1/c^2)\,\partial^2\psi'/\partial T^2 = 0$$

The two forms express the same propagation speed after conversion between temporal and spatial units.

where:

ψ, ψ'	–	Time-primary and space-primary field descriptions.
T	–	Temporal coordinate.
S	–	Spatial coordinate.
c	–	Vacuum light speed.

4.2. Energetic Exclusivity

$$E(T, S) = E_0 \exp(\alpha T + \beta S)$$

where:

$E(T,S)$	–	Energy as a function of temporal and spatial coordinates.
E_0	–	Positive reference energy.
α	–	Temporal scaling coefficient (inverse time).
β	–	Spatial scaling coefficient (inverse length).

The exponent is dimensionless. This model diverges when $\alpha T + \beta S$ tends to positive infinity. BUFT postulates the exclusion of simultaneous dimensional fixation at finite T and S.

5. BUFT Theorem I: Duality of Dimensional Genesis

STATEMENT

At the moment of cosmogenic origin (T = 0), the perceived emergence of the universe is governed by the observer's fixed dimensional reference. An observer fixed in time witnesses spatial branching constituting a multiverse. An observer fixed in space perceives temporal branching constituting a megaverse. BUFT postulates that these frames are physically valid and that simultaneous stationary observation of both dimensions requires infinite energy. The following argument establishes the stated mathematical duality conditional on these postulates.

PROOF SKETCH

1. Let $\psi(T, S) = \tau(T) \cdot \sigma(S)$ be the BEF state. Normalize its coordinates as in section 3.2.

2. Apply the linear involution:

$$D\psi(\theta, \xi) = \psi(\xi, \theta); \quad D^2 = I.$$

3. For wave-equation solutions, use the two frame forms:

$$\partial^2\psi/\partial T^2 - c^2\,\partial^2\psi/\partial S^2 = 0$$

$$\partial^2\psi'/\partial S^2 - (1/c^2)\,\partial^2\psi'/\partial T^2 = 0$$

4. $E(T, S) = E_0 \exp(\alpha T + \beta S)$ diverges only as its exponent tends to positive infinity. The BUFT postulate excludes finite-coordinate dual fixation.

6. Implications and Applications

6.1. Relativistic Cosmology

BUFT's dual-frame genesis introduces a transformative perspective on cosmic evolution. It yields unified explanations for the inflationary epoch, Cosmic Microwave Background anisotropy, and the genesis of vacuum fluctuations.

Inflationary Epoch and Early Universe Topology:

The dual-frame model—*distinguishing between a time-primary observer's multiverse, and a space-primary observer's megaverse*—suggests that early-universe topology and entropy flow are not singular phenomena but complementary outcomes, dependent on the observer's dimensional orientation. This duality implies that the initial cosmic expansion may be interpreted either as a rapid spatial divergence (multiverse) or as a temporal bifurcation (megaverse). Both are structurally equivalent under BUFT.

Cosmic Microwave Background Anisotropy:

Anisotropies in the CMB may be viewed as signatures of Time-Force fluctuations within the Bakous Energy Field (BEF). In this framework, gravity emerges from fluctuations in the Time-Force, which govern matter-energy interactions and the universe's dynamic curvature. Consequently, the observed CMB variations encode information about the primordial timespace curvature and the initial conditions of BEF resonance.

Vacuum Fluctuation Genesis:

BUFT reinterprets the quantum genesis at $T = 0$ through observer-based dimensional orientation. Fluctuations in the BEF are not random quantum events but structured, time-force-driven deformations of the nascent curvature field. These form the gravitational and topological underpinnings of the emergent universe.

6.2. Quantum Interpretation

Within BUFT, quantum measurement transcends statistical abstraction, emerging instead as a direct result of dimensional collapse and observer fixation-bringing quantum uncertainty into direct alignment with field dynamics.

Dimensional Collapse and Probabilistic Branching:

Measurement is interpreted as a collapse of higher-dimensional possibility into observer-constrained actuality. Fixation in one primary dimension (either time or space) induces probabilistic branching in the other. This branching is not arbitrary but guided by the resonance patterns within the BEF. Conscious observation engages this field, shaping the realized quantum state by modulating its energetic curvature.

Reconciling Everettian Multiverse and Entangled Decoherence:

BUFT offers a mechanism to reconcile the Everettian view with decoherence theory. The observer's gravitational-temporal frame breaks symmetry within the BEF, selecting a resonant eigenstate from a spectrum of latent configurations. The wavefunction thus becomes not a cloud of potential but a projection of a coherence manifold intersected by conscious measurement-a dynamic equilibrium stabilized through observation.

6.3. Consciousness and Field Dynamics

BUFT elevates consciousness from an emergent property of neurobiology to a foundational dimension of the universe—a fifth-dimensional interface essential to the structure of reality itself.

Observational Anchoring and BEF Modulation:

The act of observational anchoring-where consciousness localizes itself in a specific dimensional orientation—directly interacts with and modulates the BEF. Through this mechanism, consciousness doesn't merely perceive reality but co-constructs it. Physical outcomes are shaped by the curvature response of the BEF to intentional awareness.

Echoes of Quantum Contextuality:

This principle resonates with the concept of quantum contextuality, in which measurement outcomes depend on the broader experimental configuration. BUFT extends this to include observer intent and dimensional fixation, embedding consciousness as a variable within the measurement process itself.

Support for Participatory Cosmology:

By formalizing consciousness as a field-interactive dimension, BUFT provides robust theoretical grounding for participatory cosmology. In this framework, observers are not passive entities but active agents of spacetime modulation. Measurement is redefined as a local resonance stabilization event within the BEF, conditioned by curvature, dimensional fixation, and gravitational coherence.

7. Future Work

Future directions of research under the BUFT framework include the derivation of quantum field equations that govern **Bakous Chirality Nodes** (BCNs), which replace conventional excitons and serve as the fundamental quanta in timespace interactions. Additionally, deeper mathematical exploration of Time-Force curvature metrics will be pursued to describe how temporal dynamics directly modulate the geometry of space through BEF fluctuations. Another avenue of development involves correlating these metrics with observable cosmological phenomena, such as redshift asymmetries and inflation phase transitions, to establish empirical validation. Finally, simulations of dual-frame bifurcation dynamics in controlled quantum computing platforms may offer practical demonstrations of observer-dependent cosmogenesis. As a result, it may provide further insight into the operational structure of multiversal and megaversal evolution within the BUFT paradigm.

References

1. Bakous, R. (2025). *Bakous Unified Field Theory: A Timespace Perspective.* Independently Published.
2. Misner, C., Thorne, K., & Wheeler, J. (1973). *Gravitation.* W.H. Freeman.
3. Hawking, S. (1988). *A Brief History of Time.* Bantam.
4. Carroll, S. (2010). *From Eternity to Here: The Quest for the Ultimate Theory of Time.* Dutton.
5. Penrose, R. (2004). *The Road to Reality.* Jonathan Cape.
6. Everett, H. (1957). *"Relative State" Formulation of Quantum Mechanics.* Rev. Mod. Phys.
7. Tegmark, M. (2003). *Parallel Universes.* Scientific American, 288(5), 40-51.

—※—※—※—

ARTICLE 4

The Eight Dimensions of Time-Space in the Bakous Unified Field Theory

ABSTRACT

This paper presents a formal exposition of the eight dimensions constituting timespace under the Bakous Unified Field Theory (BUFT). Unlike traditional models that treat spacetime as a four-dimensional manifold, BUFT proposes a fundamentally distinct structure in which time precedes space ontologically, and the fabric of reality emerges from energy-matter fluctuations within the Bakous Energy Field ($\mathfrak{B}$). Each of the eight dimensions plays a discrete but interdependent role in structuring reality, from spatial extension to consciousness, entropy, and the governing laws of physics. The paper details the operational characteristics, inter-dimensional dependencies, and mathematical underpinnings that allow these eight dimensions to integrate classical, quantum, and relativistic physics into a unified framework.

1. Introduction

Conventional physics distinguishes space and time as coequal components of a four-dimensional manifold. The Bakous Unified Field

Theory (BUFT), however, redefines this foundation by introducing **timespace**—a unidirectional continuum emergent from oscillatory behavior in the Bakous Energy Field ($\mathfrak{B}$), which itself is the primordial substrate of all physical structure. This reconceptualization allows BUFT to resolve long-standing incompatibilities between general relativity, quantum mechanics, and string theory.

At the heart of BUFT lies the **Bakous Energy Field ($\mathfrak{B}$)**, an omnipresent, pre-material energetic medium that undergoes harmonic oscillations, giving rise to matter, forces, and space. The eight dimensions of timespace are hierarchically structured, with the first three being geometric, and the remaining five encoding the dynamical, informational, and metaphysical strata of physical law.

2. The Eight Dimensions of Time-Space

Each dimension in the BUFT framework contributes uniquely to the structure, function, and causal architecture of the universe. Their definitions, operational principles, and mathematical expressions are detailed below.

2.1. First Dimension – Length

$$X(t) = X_0 \cdot e^{iwt}$$

where:

$X(t)$	–	Linear spatial extension as a function of time
X_0	–	Initial position amplitude
w	–	Angular frequency of oscillation
t	–	Time parameter

In BUFT, length defines the first visible thread of structure. It is the straight line that energy travels along when it first takes form. This is the simplest and most foundational spatial extension.

2.2. Second Dimension — Width

$$x_2 \perp x_1$$

where:

x_2	—	Width coordinate along the second axis.
$\perp$	—	Orthogonality relation between the indicated axes.
x_1	—	Length coordinate along the first axis.

Width builds on length by adding a second direction, letting things spread out into surfaces. It allows fields to stretch, bend, and ripple beyond a single line, forming the basis for planes and wavefronts.

2.3. Third Dimension — Depth

$$\mathrm{X} = x_1\mathbf{e}_1 + x_2\mathbf{e}_2 + x_3\mathbf{e}_3$$

where:

X	—	Position vector with components (x_1, x_2, x_3).
x_1	—	Length coordinate along the first axis.
x_2	—	Width coordinate along the second axis.
x_3	—	Depth coordinate along the third axis.
$\mathbf{e}_1$	—	Dimensionless basis vector along the first axis.
$\mathbf{e}_2$	—	Dimensionless basis vector along the second axis.
$\mathbf{e}_3$	—	Dimensionless basis vector along the third axis.

Depth completes physical space as we experience it. With this third direction, objects can fully form, take up volume, and exist in the physical world with boundaries and enclosures.

2.4. Fourth Dimension — Time

$E(t) = mc^2 \cdot e^{t/\tau}$

where:

$E(t)$ — Energy as a function of temporal amplification

m — Rest mass of the system

c — Speed of light

t — Temporal coordinate

τ — Characteristic timescale of temporal energy growth.

Time in BUFT is not just a passive dimension but an active force. It is the first emergent property of the Bakous Energy Field and serves as the origin of direction, change, and causality. Time governs the rate at which reality unfolds and links all events into an irreversible continuum. Unlike coordinate time in relativity, BUFT's time is a field that curves, expands, and pressures the formation of matter and structure through the time-force dynamic.

2.5. Fifth Dimension — Consciousness

$O_c = E_{em} \cdot e^{-\beta\chi}$

where:

O_c — Energy-valued influence of conscious observation.

E_{em} — Electromagnetic energy of neural emissions

β — Cognitive coefficient ensuring $\beta\chi$ is dimensionless.

χ — Consciousness field variable

Consciousness in BUFT is not a byproduct of biology but a core dimension of the universe. It directly interacts with the Bakous Energy Field and alters its configuration. Through electromagnetic emissions and coherent thought structures, consciousness collapses quantum fields and selects between possible states. It introduces intentionality and awareness into physical law, ensuring that observation plays a formative role in shaping the outcomes of quantum and macroscopic systems.

2.6. Sixth Dimension – Entropy

$$S_{\mathrm{BUFT}} = \eta \cdot \frac{\mathrm{d}\mathcal{T}}{\mathrm{d}\tau} \cdot \frac{1}{|\nabla \cdot \mathcal{B}|}$$

where:

S_{BUFT}	– Entropy as a dimensionally propagating field in timespace.
η	– Entropic encoding coefficient governed by the BEF.
$\frac{\mathrm{d}\mathcal{T}}{\mathrm{d}\tau}$	– Rate of time-force change with respect to proper time τ.
$\nabla \cdot \mathcal{B}$	– Divergence of the Bakous Energy Field; its magnitude scales directly with gravitational intensity in this relation.

For fixed η and temporal-rate factor, the displayed relation therefore makes entropy inversely proportional to gravitational intensity. It applies only where $|\nabla \cdot \mathcal{B}| > 0$.

Entropy in BUFT is the sixth spatial dimension—no longer a scalar statistic, but a curvature-inverting geometric field that operates across timespace. It functions as a dynamic system of energy redistribution, ensuring continuity of structure across collapse events, observational thresholds, and curvature gradients. Entropy records conscious interaction into the topology of the universe and upholds the conservation of information across gravitational and quantum boundaries. It is the geometric implementation of the laws of thermodynamics, embedded in the evolving architecture of the cosmos.

2.7. Seventh Dimension — Bakous Energy Field

$$E_{\mathscr{B}} = E_{em} + E_g + E_t$$

where:

$E_{\mathscr{B}}$ – Total energy within the Bakous Energy Field

E_{em} – Electromagnetic energy

E_g – Gravitational energy

E_t – Temporal energy component

The Bakous Energy Field is the source layer of reality. It's a hidden medium that vibrates and fluctuates, generating all physical phenomena. Everything in the universe is born from it.

2.8. Eighth Dimension — Governing Laws of Physics

$$\Lambda \equiv Constancy\ of\ Physical\ Law\ across\ Time - Space$$

where:

Λ – Governing framework for symmetries, invariants, and constants

Encodes consistency in all interactions and dimensional transitions

The eighth dimension holds the blueprint. It contains the invisible code that tells energy how to behave, how matter should form, and how timespace must evolve. It is the law behind all laws.

3. Interdimensional Coupling and Mathematical Cohesion

The eight-dimensional timespace continuum is not merely a stacked hierarchy, but a dynamically coupled manifold. Interdimensional dependencies preserve coherence through recursive symmetry and conservation laws.

$$\nabla \cdot \mathscr{F}(x_1, x_2, x_3, t, \chi, S, \mathscr{B}, \Lambda) = 0$$

where:

$\nabla \cdot \mathscr{F}$ — Divergence of the composite field in the stated constraint.

$\mathscr{F}$ — Composite field representing coupled dimensional dynamics

Supporting harmonic expressions:

$$X(t) = X_0 \cdot e^{iwt}$$

$$S \propto \frac{d\mathscr{T}}{d\tau} \cdot \frac{1}{|\nabla \cdot \mathscr{B}|}$$

$$E(t) = mc^2 \cdot e^{t/\tau}$$

For fixed encoding and temporal-rate factors, entropy varies inversely with gravitational intensity through the divergence magnitude. These expressions require $|\nabla \cdot \mathscr{B}| > 0$.

4. Implications for Physical Theory

BUFT redefines the fundamental structure of reality by embedding known physics within a deeper dimensional architecture. Gravity becomes the visible effect of entropy's curvature over timespace, not merely mass-attraction. Cosmic expansion arises not from a separate dark energy but from the exponential propagation of time-force within the Bakous Energy Field. Consciousness is reframed as a dimension that participates in physics by selecting outcomes, leaving informational residues on timespace geometry.

Thermodynamics is no longer statistical but geometric-entropy becomes a field-like process that balances structure against deformation. Quantum effects emerge naturally as harmonic modes of $\mathfrak{B}$, while classical forces represent stable oscillatory regimes. The laws of physics themselves are revealed as expressions of the eighth dimension, ensuring coherence and invariance across the evolving timespace manifold. BUFT thus unifies the cosmos not just mathematically, but causally, structurally, and philosophically.

5. Conclusion

The eight-dimensional timespace model in BUFT unifies the domains of quantum mechanics, relativity, and cosmology under a single ontological and mathematical structure. Each dimension defines an essential pillar of physical reality, harmonized through recursive dynamics, entropy-regulated curvature, and observer-activated phenomena. The theory provides a comprehensive account of the universe's architecture and evolution.

References

1. Einstein, A. (1905). *Zur Elektrodynamik bewegter Karper.* Annalen der Physik.
2. Planck, M. (1900). *On the Theory of Energy Radiation.* Annalen der Physik.
3. Wheeler, J. A. (1962). *Geometrodynamics.* Academic Press.

—※—※—※—

Article 5

The Bakous Unified Field Theory (BUFT): A Framework Unifying Gravity, Quantum Mechanics, Classical Mechanics, and String Theory

Abstract

The Bakous Unified Field Theory (BUFT) represents a groundbreaking paradigm in theoretical physics, integrating gravity, quantum mechanics, classical mechanics, and string theory under a unified framework. BUFT redefines the relationship between time and space—termed here as 'timespace'—and postulates an initial quantum fluctuation, with energy propagating at light speed. The theory introduces the Bakous

Energy Field (BEF) as the underlying field driving the expansion of timespace and unifying fundamental forces. This paper outlines the mathematical foundations, empirical connections, and predictive models of BUFT while introducing testable predictions that expand our understanding of the universe.

1. Introduction

Modern physics, despite significant advancements, remains fragmented in reconciling the fundamental forces and theories of the universe. The Bakous Unified Field Theory (BUFT) addresses this gap by unifying these domains under the framework of timespace. In the BUFT framework, temporal structure serves as the governing reference through which spatial organization emerges. BUFT postulates an initial quantum fluctuation in a one-dimensional energy and matter state, with energy propagating at light speed. The BEF precedes matter and spatial manifestation; time coexists with the BEF.

2. Foundational Framework

At $T = 0$, the universe was initiated through the interplay of pure energy and matter states via the mechanism of $E = mc^2$, triggered by a singularity originating from another universe. This quantum fluctuation, facilitated by the Bakous Energy Field, produced harmonic oscillating strings that generated timespace and matter through energy propagation at the speed of light. Each iteration of $E = mc^2$ expanded timespace exponentially, with the BEF maintaining the observed physical constants.

The Bakous Energy Field is defined as a universal, omnipresent field in which energy and matter emerge, interact, and propagate. It underpins all fundamental forces and serves as the medium for oscillations that drive the expansion of timespace.

3. Mathematical Foundations of BUFT

3.1. Time-Space Curvature and Gravity

BUFT redefines gravity as the propagation mechanism of time in timespace, with the **Bakous Energy Field (BEF)** as the foundation of gravitational dynamics. The gravitational field, inversely proportional to the magnetic field, is expressed through modified equations of timespace curvature.

BUFT proposes the following curvature relation; its application requires a tensor-consistent, dimensionally normalized coupling:

$$c \dagger t_s = \frac{8\pi G}{c^4} T_{\mu\nu} + \frac{1}{\mu_0 c^2} (\nabla \times B)^2$$

where:

$T_{\mu\nu}$ — is the stress-energy tensor,

μ_0 — is the permeability of free space.

The BEF links gravitational and electromagnetic dynamics. Entropy scales inversely with gravitational intensity at fixed encoding and temporal-rate factors.

3.2. Electromagnetic and Gravitational Forces

BUFT posits that magnetic dipole fields, with inverse-cube distance scaling, participate in generating inverse-square gravitational effects. The BEF is the medium through which electromagnetic oscillations generate gravitational interactions. The proposed Maxwell-BUFT relations are:

$$\nabla \cdot E = \frac{\rho}{\epsilon_0}, \quad \nabla \cdot B = 0, \quad \nabla \times E = -\frac{\partial B}{\partial t}$$

$$\nabla \times B = \mu_0 J + \frac{1}{c^2}\frac{\partial E}{\partial t} - \nabla^2 \mathcal{G}$$

where:

$\mathcal{G}$ — represents the gravitational field generated by oscillating electromagnetic fields within the BEF. The coupling normalization sets the scale of the added field term.

4. Empirical Connections

4.1. Cosmic Microwave Background (CMB) Analysis

Observations of the CMB reveal harmonic oscillations consistent with the propagation of strings postulated by BUFT. These oscillations are interpreted within BUFT as remnants of $E = mc^2$ iterations in the BEF and its proposed exponential timespace expansion.

4.2. Quantum Vacuum Fluctuations

The Casimir effect provides an empirical connection to quantum vacuum phenomena. BUFT interprets these phenomena through harmonic oscillating strings in the BEF, supporting the emergence of energy and matter. Independent experiments test this BUFT interpretation.

4.3. Photoelectric Effect

The maximum photoelectron kinetic energy equals the photon energy minus the work function. BUFT proposes an additional curvature-dependent BEF contribution. The BUFT energy model is:

$$E_k = hf - \phi + \int_0^{\infty} Cts\, dx$$

where:

ϕ — is the work function,

Cts — proposed BEF curvature correction per unit length; the integral must converge to a finite energy.

5. Predictions of BUFT

5.1. Time-Space Waves

BUFT predicts that time itself exhibits wave-like properties, allowing for kinetic energy transfer in timespace. These waves propagate through the Bakous Energy Field, and their detection can be experimentally tested through high-precision interferometry, which is capable of identifying temporal distortions.

5.2. Gravitational-Magnetic Interactions

The inverse proportionality of gravitational and magnetic fields within the BEF suggests measurable magnetic-flux variations associated with gravitational waves. Testing this proposal would require magnetic-flux sensors alongside gravitational-wave interferometers.

5.3. Sleep-State Consciousness

BUFT predicts that during sleep, the brain enters an electromagnetic "stationary" state at light speed within the BEF, making sleep instantaneous from the observer's perspective. EEG and fMRI studies could explore this hypothesis by measuring changes in neural electromagnetic activity.

6. Furthering Our Understanding of the Universe

BUFT redefines fundamental physical constants as emergent properties of $E = mc^2$ iterations within the Bakous Energy Field. This provides insights into:

1. Entropy and Gravity: Entropy scales inversely with gravity at fixed encoding and temporal-rate factors.
2. **Dark Energy and Matter**: Harmonic oscillations of strings in the BEF are linked to the universe's accelerating expansion.
3. Consciousness as a Fifth Dimension: BUFT postulates interaction with the BEF through electromagnetic activity. A measurable signature remains to be established.

7. Conclusion

The Bakous Unified Field Theory unifies the core principles of modern physics into a cohesive framework, offering testable predictions and mathematical foundations. By redefining the interplay of timespace, gravity, and electromagnetism through the Bakous Energy Field, BUFT bridges the gaps between quantum mechanics, classical mechanics, and string theory.

References

1. Einstein, A. (1905). Zur Elektrodynamik bewegter Körper. *Annalen der Physik.*
2. Planck, M. (1900). *On the Theory of Energy Distribution in the Normal Spectrum.*
3. Wheeler, J. A. (1962). *Geometrodynamics.* Academic Press.
4. Hawking, S. (1988). *A Brief History of Time.* Bantam Books.

—※—※—※—

Article 6

Energy Distribution and Harmonic Oscillations of Strings through the Framework of the Bakous Unified Field Theory

Abstract

This paper delves into the profound implications of energy distribution along harmonically oscillating strings, contextualized within the Bakous Unified Field Theory (BUFT). The universe's emergence through a quantum fluctuation is examined in detail, proposing simultaneous energy and matter manifestation, with energy propagation through the BEF at the speed of light.
The rest-energy relation $E = mc^2$ connects mass and energy. Within BUFT, harmonically oscillating strings propagate energy, generate space, and maintain universal constants. The interplay between gravity, electromagnetism, and timespace curvature is analyzed to highlight the predictive framework of BUFT.

Furthermore, mathematical and experimental insights into oscillating strings motivate the proposed BUFT account of timespace structure and cosmic energy distribution.

1. Introduction

The Bakous Unified Field Theory (BUFT) provides a novel framework for understanding the origins and dynamics of the universe. According to BUFT, the universe began through a quantum fluctuation, wherein pure energy and matter co-existed in a one-dimensional state, with energy propagating at the speed of light. The rest-energy relation $E = mc^2$ connects mass and energy; BUFT describes the initial state and spatial formation through BEF dynamics. Time coexists with the BEF.

The BEF functions as the pre-material energetic substrate through which harmonic propagation distributes energy into observable structure, an essential aspect in the propagation and generation of space. Harmonic oscillations in strings, as described by BUFT, are central to the propagation of energy and the continuous generation of space. These oscillations, which occur at relativistic speeds, maintain the constancy of universal laws and ensure the preservation of fundamental constants.

This paper investigates how oscillating strings operate within this framework, continuously maintaining the equilibrium of universal constants by adhering to the principles of $E = mc^2$. Detailed mathematical models will illustrate the interaction between the Bakous Energy Field and oscillating strings, along with the effects of this interaction on timespace curvature at different observational scales.

2. The Role of $E = mc^2$ in Energy Distribution

The rest-energy relation $E = mc^2$ forms a cornerstone of BUFT. It relates rest mass to energy; the field dynamics govern conversion rates and oscillations. BUFT postulates continuous energy propagation through the BEF at the speed of light. Energy transferred into strings and matter must be accounted for together with the energy supplied by the BEF.

MATHEMATICAL REPRESENTATION OF ENERGY IN OSCILLATING STRINGS:

In an effective small-amplitude transverse-string model, the integral over string length L gives total oscillation energy. Its integrand is energy per unit length:

$$E = \int_0^L \left[\frac{1}{2}\mu\left(\frac{\partial y}{\partial t}\right)^2 + \frac{1}{2}T\left(\frac{\partial y}{\partial x}\right)^2\right]dx$$

where:

E – represents the total oscillation energy of the string.

μ – is the mass per unit length of the string.

T – String tension governing the transverse restoring force.

y(x, t) – is the displacement of the string as a function of position *(x)* and time *(t)*.

For uniform tension and mass density, $v^2 = T/\mu$. Propagation at c requires $T = \mu c^2$. The quantity c^2 is a squared speed, not energy.

As the string oscillates, its local energy density varies with position and time. BUFT associates energy propagation through the BEF with the continued generation of space.

3. Harmonic Oscillations and Space Generation

Harmonic oscillations of strings play a fundamental role in BUFT's proposed generation and expansion of space. Energy supplied through the BEF propagates outward at the speed of light, contributing to spatial expansion within

BUFT. It ensures that space continues to grow and evolve as energy is transferred from one oscillation to the next.

DETAILED MECHANISM OF SPACE GENERATION:

1. Relativistic Oscillations and Energy Propagation:

The oscillating strings propagate energy as waves at the speed of light. These oscillations transfer energy through the Bakous Energy Field. BUFT postulates that each oscillation produces ripples in space and contributes to new spatial extent. This continuous generation of space is a key mechanism for the expansion of the universe.

2. Energy Distribution and Continuous Expansion:

As the oscillating strings move through space, they create an exponential expansion of space itself. This process is self-sustaining; each oscillation generates more space, which then facilitates the propagation of subsequent oscillations. This mechanism aligns with BUFT's prediction of an exponentially expanding universe driven by the harmonic oscillations of strings.

3. Wave-Like Properties of Strings:

The oscillations of strings exhibit wave-like properties, meaning that they propagate energy in the form of waves that interact with the Bakous Energy Field. This interaction leads to the continuous creation of space and energy distribution at relativistic speeds. As the strings oscillate, they ensure that the speed of light remains constant across all scales, preserving the integrity of universal constants.

4. Time-Space and Its Interaction with Oscillating Strings

Time-space is the fundamental fabric of the universe, governing both the progression of time and the structure of space. The interaction between timespace and oscillating strings is central to understanding the dynamics of the universe, especially as described by BUFT. The Bakous Energy Field serves as the medium through which oscillations occur, generating space and maintaining the constancy of time.

PRE-INTERACTION PHASE:

Before the oscillations of strings begin, the Bakous Energy Field exists in a state of pure potential energy, unmanifested as matter but brimming with the capacity to facilitate the conversion of energy into mass. In this phase, the field remains unexcited as a pre-spatial energetic substrate. A spatial energy distribution is defined only after space emerges. The BEF supplies energy for subsequent interactions, with transfers accounted for in the total energy balance.

DURING THE INTERACTION WITH OSCILLATING STRINGS:

As the strings begin to oscillate, they interact with the Bakous Energy Field, transferring energy through the field and creating ripples in space. The oscillations propagate through the field at the speed of light, with the energy in the strings contributing to the expansion and generation of new space. This interaction with the Bakous Energy Field is essential in maintaining the propagation of energy and the creation of mass, as described by $E = mc^2$.

During this phase, the Bakous Energy Field is visibly disturbed by the oscillations. These oscillations are not only responsible for energy transfer but also for the continuous modification of the structure of space itself. As each oscillation passes through the BEF, it contributes to the ongoing expansion of the universe, increasing spatial extent and further influencing the curvature of timespace.

POST-INTERACTION PHASE:

After the oscillations have moved through the Bakous Energy Field, the field continues to propagate the energy throughout space. This post-interaction phase ensures that the Bakous Energy Field remains active, continuing to fuel spatial expansion through energy supplied by the BEF. The field, though disturbed during the oscillations, returns to a state of dynamic equilibrium, where it facilitates the ongoing creation of mass and the preservation of universal constants. This continuous flow of energy within the BEF is responsible for maintaining the integrity of timespace and the fundamental laws of the universe.

The BEF, therefore, plays an essential role in the ongoing dynamics of the universe. It ensures that energy remains in flux while space continues to expand. This interaction between the BEF and the oscillating strings is a crucial aspect of the overall framework of BUFT, illustrating the continual relationship between energy, mass, and space.

5. Conclusion

The Bakous Unified Field Theory offers a comprehensive understanding of the continuous interaction between harmonically oscillating strings and the BEF. BUFT postulates that these oscillations preserve universal laws and facilitate continuous space generation and energy transfer through the field. The rest-energy relation $E = mc^2$ connects mass and energy; the string equations govern motion. This framework offers a proposed account of cosmic evolution across microscopic and macroscopic scales.

References

1. Einstein, A. (1905). Does the Inertia of a Body Depend Upon Its Energy Content? *Ann. Phys.*, 18, 639-641.
2. Planck, M. (1900). *Zur Theorie des Gesetzes der Energieverteilung im Normalspektrum. Verh. Dtsch. Phys. Ges.*, 2, 237-245.
3. Green, M. B., Schwarz, J. H., & Witten, E. (1987). *Superstring Theory* (Vols. 1-2). Cambridge Univ. Press.
4. Misner, C. W., Thorne, K. S., & Wheeler, J. A. (1973). *Gravitation.* W.H. Freeman.
5. Carroll, S. M. (2004). *Spacetime and Geometry: An Introduction to General Relativity.* Addison-Wesley.
6. Zwiebach, B. (2004). *A First Course in String Theory.* Cambridge Univ. Press.

—※—※—※—

ARTICLE 7

The Role of Time-Force Fluctuations in the Initiation of the Universe: A Bakous Unified Field Theory Perspective

ABSTRACT

This paper investigates the hypothesis that time-force fluctuations, mediated by the Bakous Energy Field (BEF), are instrumental in the formation of the universe. These fluctuations lead to the emergence of particles with opposing charges governed by the electroweak force, whose interactions release energy that is subsequently converted into matter. The Bakous Unified Field Theory (BUFT) integrates gravity and quantum forces into a single framework, offering a new perspective on the origin and evolution of the universe.

1. Introduction

The origins of the universe remain one of the most profound mysteries in physics. The hot Big Bang model describes expansion from an early hot, dense state; it does not establish the universe's ultimate origin. The Bakous Unified Field

Theory (BUFT) proposes that time-force fluctuations, driven by the Bakous

Energy Field (BEF), are responsible for the creation of particles. These fluctuations set in motion processes that lead to the gradual condensation of energy, transforming it from a highly energetic, formless state into distinct particles and, eventually, matter. This transition marked the beginning of the universe's structural formation, shaping the fundamental building blocks from pure energy. This paper explores how the Bakous Unified Field Theory (BUFT) offers a unified framework for understanding the cosmic origins and the mechanisms that lead to the emergence of matter.

2. The Bakous Unified Field Theory (BUFT)

At its essence, the Bakous Unified Field Theory (BUFT) provides a unified view of forces and particles through fluctuations of the BEF, from which spatial structure emerges. Central to this theory is the concept of time-force, a dynamic field that governs the emergence of gravitational, electromagnetic, weak, and strong nuclear forces. Standard electroweak theory already unifies electromagnetic and weak interactions. BUFT proposes a deeper unification of all forces through fluctuations of the BEF. This field is understood as a medium through

which force quanta, such as **Thoton** (time propagation), **Photon** (electromagnetic interaction), and **Gravistar** (gravitational interaction), mediated these forces. In this framework, the time-force acted not as a passive background but as an active, shaping force that influenced the behavior of the universe at its most fundamental level. This provided a novel perspective on the origin of the universe and the processes by which matter emerged from pure energy.

3. Time-Force and the Transformation of Energy into Matter

In the very early moments of the universe, when energy existed in a highly concentrated and undifferentiated form, time-force fluctuations mediated by the Bakous Energy Field (BEF) initiated the transformation of this energy into distinct particles. These fluctuations set the stage for the emergence of fundamental particles from the raw quantum state. As the energy condensed, a diverse array of particles began to form, including quarks, leptons, and photons. Each of these particles interacted with the fabric of timespace, with their properties being shaped by the fluctuations in the time-force field. Quarks and leptons, for example, interacted with each other through the fundamental forces governed by time-force, leading to the creation of more complex structures. These interactions also created disturbances in the surrounding timespace, further influencing the movement and behavior of these particles.

As these particles continued to form and interact, they began to combine in specific configurations that gave rise to more complex structures, laying the groundwork for the eventual emergence of atoms. Through time-force fluctuations, the raw energy from the early universe was progressively condensed into the building blocks of matter, each particle playing a unique role in shaping the cosmos. The dynamic relationship between time-force and these particles was essential in driving the early stages of cosmic evolution, ensuring that the transition from pure energy to matter was not only possible but inevitable.

4. The Electroweak Force and Early Particle Interactions

Electroweak interactions participated alongside other fundamental interactions in the early universe. BUFT associates the particles involved with time-force fluctuations. As the universe cooled, electroweak symmetry breaking gave rise to distinct electromagnetic and weak interactions. Charged-particle pairs are not created solely by this symmetry breaking. The interactions of electrons and positrons include pair creation and annihilation, subject to energy, momentum, and charge conservation. These processes contributed to the early evolution of the universe.

In standard electroweak theory, symmetry breaking through the Higgs field generates masses for the W and Z bosons; quarks and charged leptons acquire masses through their couplings to that field. The photon remains massless. Within BUFT, the Bakous Energy Field is proposed as the underlying substrate through which particles and mass emerge. BUFT interprets this process through temporal-field dynamics. Comparison with the Higgs mechanism concerns the correspondence between the field quantities. As the universe evolved and cooled, strong and electromagnetic interactions also played essential roles in forming nuclei, atoms, and larger structures. BUFT places these processes within its time-force framework.

5. Conclusion

The investigation into the very early moments of the universe through the lens of the Bakous Unified Field Theory (BUFT) reveals a highly intricate and dynamic process governed by time-force fluctuations that led to the formation of the building blocks of matter. The transformation of energy into particles, facilitated by the Bakous Energy Field (BEF), offers a new understanding of the way fundamental forces were interwoven in the universe's inception. The interactions between opposing particles, regulated by the electroweak force, reveal a profound connection between the forces that shaped the early universe.

Drawing insights from this framework, we see that time-force is not merely a passive backdrop but a central player in the creation and evolution of the cosmos. As time-force fluctuations led to the condensation of energy into distinct particles, these particles engaged in interactions that, over time, gave rise to the stable structures that formed the foundation for everything we observe today. The transformation of energy into matter was not a simple process but one that involved delicate, ongoing interactions within the fabric of timespace, culminating in the formation of the universe's earliest structures. By integrating all forces under a unified framework, the Bakous Unified Field Theory (BUFT) provides a compelling new perspective on the origins of the universe and its continuing evolution.

References

1. Planck, M. (1901). *Ueber das Gesetz der Energieverteilung im Normalspectrum.* Annalen der Physik, 309(3), 553-563.
2. Einstein, A. (1905). *Does the Inertia of a Body Depend Upon Its Energy Content?* Annalen der Physik, 18, 639-641.
3. Heisenberg, W. (1927). *The Physical Content of Quantum Kinematics and Mechanics.* Zeitschrift *für* Physik, 43, 172-198.
4. Glashow, S. L. (1961). *Partial Symmetries of Weak Interactions.* Nuclear Physics, 22, 579-588.
5. Weinberg, S. (1967). *A Model of Leptons.* Physical Review Letters, 19(21), 1264-1266.
6. Salam, A. (1968). *Weak and Electromagnetic Interactions.* In N. Svartholm (Ed.), *Elementary Particle Theory: Relativistic Groups and Analyticity* (Nobel Symposium No. 8, pp. 367-377). Stockholm: Almqvist & Wiksell.
7. Englert, F., & Brout, R. (1964). *Broken Symmetry and the Mass of Gauge Vector Mesons.* Physical Review Letters, 13, 321-323.
8. Higgs, P. W. (1964). *Broken Symmetries and the Masses of Gauge Bosons.* Physical Review Letters, 13, 508-509.
9. Carroll, S. M. (2004). *Spacetime and Geometry: An Introduction to General Relativity.* Addison-Wesley.

—※—※—※—

ARTICLE 8

Time-Force Unification: The Emergence of Fundamental Forces and Particles from Temporal Dynamics

ABSTRACT

This paper introduces a unifying theoretical framework under the Bakous Unified Field Theory (BUFT), proposing that all fundamental forces and particles emerge as manifestations of time-force interactions. Time-space fluctuations govern the propagation, structure, and interaction of force-carrying quanta, establishing time-force as the foundational principle of physical reality. The quanta and states—Photon, Choton, Thoton, Gravistar, Tochon, and Trion—are localized timespace perturbations. Their properties arise from the interplay between electromagnetic, gravitational, and quantum energy densities.

In this paradigm, gravity is not an independent force but a consequence of time-force modulation, while electromagnetism, the weak force, and the strong nuclear force emerge as secondary effects of this underlying temporal field. Conscious observation, mediated by the Choton, plays a critical role in collapsing these interactions into perceivable reality, positioning the observer as an intrinsic participant in the universe's governing dynamics. This framework offers a novel perspective on the unification of fundamental interactions, redefining conventional physical laws as emergent properties of timespace fluctuations.

1. Introduction

The interplay between fundamental forces and their force-carrying quanta has long been studied within classical and quantum field theories. However, the underlying principle connecting these forces has remained elusive. Bakous Unified Field Theory proposes that differentiated force interactions arise from asymmetrical fluctuations embedded within evolving timespace structure. Time-force governs the propagation of all fundamental particles and energy distributions, making it the primary unifying principle in physics. This framework redefines gravity as a consequence of time-force modulation, eliminating the need for separate fundamental interactions.

2. Time-Space as the Governing Medium

A central distinction within BUFT is the differentiation between spacetime and timespace. General relativity describes gravitation through the geometry of four-dimensional spacetime. BUFT instead places time-force dynamics at the foundation of spatial structure, with fluctuations governing the properties of energy and matter. Timespace is dynamic, interacting with force-carrying particles at all scales.

Fluctuations within timespace result in localized perturbations, giving rise to fundamental particles. The Thoton, responsible for time propagation, ensures that temporal evolution remains continuous. The Choton governs conscious observation, bridging timespace interactions with cognitive perception. The Photon, traditionally defined as the carrier of the electromagnetic force, is observed through Choton interactions, making it intrinsically linked to conscious perception. Gravistars, Tochons, and Trions emerge as perturbations that interact with quantum energy densities, establishing the fundamental nature of gravity and dream-state physics.

3. Gravity as a Consequence of Time-Force Modulation

In classical physics, gravity is modeled as a curvature of spacetime influenced by mass-energy distributions. BUFT reframes this notion by postulating that gravity is a consequence of time-force modulation rather than a fundamental interaction. The time-force fluctuations redistribute quantum energy densities, generating variations in gravitational fields without requiring mass-dependent curvature.

BUFT proposes the following gravitational-potential equation, with a coefficient representing time-force modulation:

$$\nabla^2 \Phi_g - \frac{1}{c^2}\frac{\partial^2 \Phi_g}{\partial t^2} = T_f \nabla^2 \Psi$$

where:

Φ_g — represents the gravitational potential,

c — is the speed of light,

t — represents the time coordinate,

T_f — is the scalar time-force modulation coefficient,

Ψ — is a scalar field describing quantum energy distributions,

∇^2 — is the spatial Laplacian operator.

The product of the modulation coefficient and quantum-energy field must have gravitational-potential units. This equation is a proposed BUFT coupling model.

4. Electromagnetism, Weak Force, and Strong Nuclear Force as Derivatives of Time-Space

The remaining fundamental interactions—electromagnetism, the weak force, and the strong nuclear force—are derivatives of timespace fluctuations rather than distinct forces. Electromagnetic interactions arise due to localized variations in timespace density, where the Choton governs their perception. The weak force, responsible for particle decay, emerges from timespace asymmetries that induce probabilistic transitions between quantum states. The strong nuclear force, traditionally modeled through quantum chromodynamics, is reinterpreted under BUFT as a secondary manifestation of timespace compression, where energy redistributions maintain nuclear stability.

5. The Role of Force-Carrying Quanta

BUFT describes eight quanta or quantum states, including the composite Gravistar, as localized timespace perturbations. Their proposed interactions govern the phenomena discussed here:

- **Photon (y)**: Governs electromagnetic interactions and is perceived through Choton interactions.
- **Choton (x)**: Governs conscious observation, determining the collapse of quantum states.
- **Thoton (t)**: Enables the propagation of time, ensuring continuity in temporal evolution.
- **Tochon (T^2)**: Defines deep sleep states, interacting with cognitive timespace fluctuations.
- **Trion (T)**: Governs dream-state physics, existing as an intermediary between perception and unconscious cognition.
- **Graviton (g)**: Mediates gravitational interaction at the quantum level, transmitting curvature effects across timespace.

- Gluon (g): Massless carrier of the strong interaction between quarks. Nuclear binding involves the residual strong interaction.
- **Gravistar (G)**: A coherence state arising from the simultaneous entanglement of thoton (t), photon (y), choton (x), graviton (g), and gluon (g) and interactions—expressing a compounded modulation of gravitational, electromagnetic, observational, and nuclear force domains within timespace.

6. Conscious Observation as an Intrinsic Factor in Universal Dynamics

The Choton plays a unique role in BUFT, as it directly governs the perception of fundamental interactions. Classical physics assumes an objective reality independent of observation, but BUFT proposes that conscious observation collapses timespace fluctuations into distinct perceivable states. This establishes a direct link between cognition and physical reality, making the observer an intrinsic factor in universal dynamics.

Quantum measurement phenomena motivate this BUFT interpretation. Within BUFT, conscious observation and Choton-mediated interactions form the proposed account of these measurement phenomena. BUFT postulates the Choton as the link between conscious observation and perceived physical states.

7. Mathematical Formalization of Time-Force Interactions

BUFT proposes a local balance equation for timespace density, its transport by a velocity field, and a source representing time-force generation:

$$\frac{\partial \rho_T}{\partial t} + \nabla \cdot (\rho_T v) = S_{T_f}$$

where:

ρ_T — represents local timespace density,

v — is the velocity vector transporting that density,

S_{T_f} — is the source rate: timespace density per unit time,

$\nabla \cdot$ — is the spatial divergence operator.

This equation balances local density change, transport, and generation. Force-specific predictions require a defined source and further dynamical relations.

8. Conclusion

The Bakous Unified Field Theory (BUFT) establishes a framework in which all fundamental forces and particles emerge as manifestations of time-force interactions. Time-space fluctuations dictate the behavior of force-carrying quanta, redefining gravity as a consequence of time-force modulation.

Electromagnetism, the weak force, and the strong nuclear force arise as secondary effects of timespace interactions, while conscious observation, governed by the Choton, defines perceived reality. This theory unifies all known interactions under a single governing principle, reshaping our understanding of fundamental physics.

References

1. Einstein A. (1916). Ann. Phys. 354(7):769-822.
2. Wheeler JA, Feynman RP. (1945). Rev. Mod. Phys. 17: 157-181.
3. Bohm D. (1952). Phys. Rev. 85:166-179.
4. Feynman RP, Leighton RB, Sands M. (1963). Feynman Lectures on Physics, Vols. I-III. Addison-Wesley.
5. Penrose R. (1989). The Emperor's New Mind. Oxford Univ. Press.
6. Rovelli C. (1996). Int. J. Theor. Phys. 35:1637-1678.
7. Barbour J. (2000). The End of Time. Oxford Univ. Press.
8. Tegmark M. (2007). In Carr B. (Ed.), Universe or Multiverse? Cambridge Univ. Press.
9. Hawking SW, Ellis GFR. (1973). Large Scale Structure of Space-Time. Cambridge Univ. Press.
10. Stapp HP. (2009). Mindful Universe. Springer.

—※—※—※—

ARTICLE 9

The Bakous Unified Field Theory: New Mathematical Foundations of Time, Gravity, and Consciousness

ABSTRACT

The Bakous Unified Field Theory (BUFT) introduces a groundbreaking approach to understanding the interplay between time, gravity, and consciousness through ten foundational equations. BUFT posits that time exhibits wave-like properties and kinetic energy, which drive its propagation via gravity. Conscious observation, linked to electromagnetism in the brain, is framed as the fifth dimension, while entropy, as the sixth dimension, is redefined as a gravitational phenomenon. The theory expands upon Einstein's mass-energy equivalence with a time-dependent exponential factor, models the expansion of timespace as a continuous process, and reinterprets fundamental forces in relation to time-force.

1. The BUFT Mass-Energy-Time Relationship

$E = mc^2 e^{\frac{t}{\tau}}$

where:

E	– BUFT energy assigned to the system in its rest frame.
m	– Fixed reference rest mass of the system.
c	– Speed of light in vacuum.
t	– Temporal progression coordinate.
τ	– Positive timescale governing the exponential energy factor.
$e^{t/\tau}$	– Dimensionless temporal amplification factor.

Description & Application:

With fixed reference mass and positive timescale, BUFT postulates exponential energy amplification from E(0) = mc^2. Any increase requires energy exchange with the BEF or a specified energy balance. BUFT describes spatial expansion through coupled field and geometric dynamics.

2. The Bakous Energy Field Interaction through E = mc²

$E_B = E_{em} + E_g + E_t$

where:

E_B	– Total energy assigned to the Bakous Energy Field in the specified physical system.
E_{em}	– Electromagnetic energy contribution associated with neural activity, light, and field interactions.
E_g	– Gravitational energy contribution to BEF dynamics.
E_t	– Temporal energy contribution associated with time-force.

Description & Application:

This relation partitions the energy assigned to the BEF into electromagnetic, gravitational, and temporal contributions for a common physical system. The BEF itself is the pre-material substrate. This identity organizes the energy contributions; field dynamics govern their evolution relative to the BEF reference at T = 0.

3. Harmonic Oscillation and Time-Space Expansion

$X(t) = X_0\, e^{i\omega t}$

where:

$X(t)$	– Complex harmonic representation of spatial displacement.
X_0	– Initial displacement amplitude.
i	– Imaginary unit, satisfying $i^2 = -1$.
ω	– Angular frequency governing the oscillation phase.
t	– Temporal progression coordinate.

Description & Application:

This expression represents harmonic oscillation; its real part gives a real displacement. With fixed amplitude and real angular frequency, its magnitude is constant. BUFT connects oscillatory BEF dynamics with space generation; the coupled dynamics determine spatial growth and the expansion rate.

4. Expansion of Time-Space through E = mc² Initiation

$$V_{time\text{-}space} \propto \sum \frac{mc^2}{\hbar}$$

where:

$V_{time\text{-}space}$	— Generated timespace volume.
m	— Mass involved in each energy initiation.
c^2	— Squared light speed; the mass-to-energy conversion factor.
$\hbar$	— Reduced Planck constant, equal to $h/(2\pi)$.

Description & Application:

The sum runs over energy initiations. Each mass-energy term divided by the reduced Planck constant has units of inverse time, not volume. A volume scaling therefore requires a proportionality factor with units m^3 s. Physical normalization sets the volume scale; the expansion dynamics determine its temporal evolution.

5. Time and Gravity: Fundamental Relationship

$$\frac{dt}{ds} = \sqrt{1 - \frac{2GM}{rc^2}}$$

where:

dt	— Proper-time increment of the static clock.
ds	— Coordinate-time increment referenced to infinity.
G	— Newtonian gravitational constant.
M	— Central gravitational mass.
r	— Schwarzschild areal radius outside the central body.

Description & Application:

This is the standard Schwarzschild clock-rate relation for a static clock outside a spherical, nonrotating source, with $r > 2GM/c^2$. Coordinate time is normalized to a clock at rest at infinity. BUFT associates gravity with forward temporal progression and distinguishes this scaling relation from the clock-rate law.

6. Modified BUFT Equation for Time-Space Interaction

$$E_t \propto \frac{c^2}{r^2} \cdot \frac{1}{\tau}$$

where:

E_t	– Time-force energy.
c^2	– Squared light speed; the mass-to-energy conversion factor.
r	– Positive radial distance from the observer.
τ	– Positive characteristic timescale.

Description & Application:

BUFT proposes inverse-square distance scaling for time-force energy. The displayed combination has units s^{-3}, so an energy relation requires a proportionality factor with units J s^3. This factor sets the physical scale of the coupling. Clock-rate changes are evaluated through the temporal dynamics.

7. Conscious Observation Electromagnetic Propagation

$O_c = E_{em} \cdot e^{-\alpha B g}$

where:

O_c	– Energy-valued proxy for conscious observation.
E_{em}	– Electromagnetic energy associated with neural activity.
B	– Magnetic-field magnitude associated with brain activity.
g	– Gravitational-field magnitude.
α	– Coupling chosen to make the exponent dimensionless.

Description & Application:

The coupling coefficient makes the exponent dimensionless. At fixed neural energy and magnetic field, a positive coupling models suppression as gravitational intensity increases. Quantitative tests of this BUFT model measure its observation coefficient and associated response.

8. Entropy as the Sixth Dimension

$S = k_B \ln\Omega \cdot \frac{g_0}{g}$

where:

S	– Energy-redistribution entropy of the modeled system.
k_B	– Boltzmann constant linking state count to entropy.
Ω	– Dimensionless number of microstates.
g	– Positive gravitational-field magnitude.
g_0	– Positive reference field normalizing gravitational intensity.

Description & Application:

At fixed microstate count and reference field, this normalized BUFT relation preserves inverse entropy–gravity scaling for positive gravitational intensity. The reference field supplies the units and fixes the physical normalization.

9. Gravitational Time-Force Interaction in High-Density Fields

$$a_T = \frac{GM}{r^2}$$

where:

a_T — Time-force-associated gravitational acceleration.

G — Newtonian gravitational constant.

M — Mass of the spherical gravitational source.

r^2 — Square of the distance from the mass center.

Description & Application:

This expression gives gravitational acceleration in the Newtonian exterior limit of a spherical source. BUFT associates this field with time-force dynamics. A force would also require a test mass. This expression is not a strong-field clock-rate law; static clock comparisons follow Section 5.

10. Electromagnetic Influence on Gravitational Time-Force Propagation

$$F_{grav} \propto E_{em} \cdot \frac{1}{r^2} \cdot \frac{c^2}{\tau}$$

where:

F_{grav} – Proposed gravitational-force magnitude.

E_{em} – Electromagnetic energy affecting the field.

r^2 – Square of the positive radial distance.

c^2 – Squared light speed; the mass-to-energy conversion factor.

τ – Positive characteristic timescale.

Description & Application:

This proposed scaling retains the electromagnetic-energy and inverse-square distance dependence. To yield force, its proportionality factor must have units s^3/m. The factor sets the force scale, while a dynamical coupling law relates the fields to temporal evolution. The expression supplies the scaling relation used in force and clock-rate modeling.

11. Conclusion

The Bakous Unified Field Theory (BUFT) proposes a mathematical framework for understanding timespace, gravity, and consciousness. Time-force dynamics connect these aspects within the BUFT interpretation. These relations distinguish energy bookkeeping and established limiting relations from proposed BUFT scaling laws. Dimensional normalization, coupled field dynamics, and empirical tests remain necessary to establish quantitative predictions.

References

1. Einstein, A. (1905). *Does Inertia Depend on Energy Content?* Ann. Phys., 18, 639-41.
2. Guth, A.H. (1981). *Inflationary Universe.* Phys. Rev. D, 23, 347-56.
3. Hawking, S. W., & Penrose, R. (1970). *Singularities in Gravitation.* Proc. Roy. Soc. A, 314, 529-48.
4. Rovelli, C. (2004). *Quantum Gravity.* Cambridge Univ. Press.
5. Penrose, R., & Hameroff, S. (1996). *Consciousness and 'Orch OR'.* J. Conscious. Stud., 3(1), 36-53.
6. Chalmers, D. J. (1996). *The Conscious Mind.* Oxford Univ. Press.
7. Smolin, L. (2013). *Time Reborn.* Houghton Mifflin Harcourt.
8. Boltzmann, L. (1877). *The 2nd Law and Probability.* Wiener Berichte, 76, 373-435.

—※—※—※—

ARTICLE 10

BOFA Tensor Equation

The BOFA tensor is expressed in components as:

$$\mathscr{B}^{\mu\nu} = \frac{\mathscr{B}_k}{\tau^o} P^{\mu} P^{\nu}$$

where:

$\mathscr{B}^{\mu\nu}$ – BOFA tensor components coupling Bakous energy, time-force dynamics, and curvature-associated timespace structure. The two free indices label components in the chosen basis.

$\mathscr{B}_k$ – Bakous energy constant setting the BEF interaction scale.

τ^o – Nonzero scalar time-force interaction factor, representing the time-force/Bakous-time ratio in BUFT. Its dimensions follow the field normalization. The superscript o is a label.

P^{μ}, P^{ν} – Components of a proposed curvature-associated timespace vector. Their vector transformation law is required for the displayed construction to define a tensor.

$\otimes$ — Tensor (outer) product. It forms a second-rank tensor from two vectors; it does not contract their component indices.

In coordinate-free form:

$$\mathcal{B} = \frac{\mathcal{B}_k}{\tau^0} P \otimes P$$

Description of the BOFA Tensor

The BOFA tensor formalism represents the proposed coupling of Bakous energy, time-force interactions, and curvature-associated structure throughout evolving timespace. Within BUFT, it expresses how the omnipresent Bakous Energy Field shapes the universe through its interaction with temporal and geometric dynamics.

The tensor product of the curvature-associated vector with itself gives the displayed component products. With a scalar energy/time-force coefficient and a vector transformation law, the result is a symmetric, factorized second-rank tensor. This construction links the geometric and energetic quantities used in BUFT's account of gravitational time dilation, energy exchange, and entropy. BUFT associates these interactions with its massless force-carrying quanta.

Each tensor component must have units consistent with the corresponding curvature-associated component product, multiplied by the Bakous energy constant and divided by the interaction factor. The field definitions and curvature measurements use a consistent physical normalization. Propagation equations and the entropy law govern this tensor's evolution.

—※—※—※—

ARTICLE 11

Validation of the Bakous Unified Field Theory: A New Perspective on Space, Time, and Causality

ABSTRACT

The Bakous Unified Field Theory (BUFT) proposes that space possesses intrinsic motion, dynamically adjusting its structure to ensure that cause-and-effect relationships remain inviolate. Moreover, when space undergoes motion, the passage of time ceases within that domain, particularly in the context of quantum interference phenomena.

This paper examines the mathematical consistency of this hypothesis, demonstrating its alignment with established principles of relativistic geometry and quantum mechanics. Furthermore, it is shown that conventional observational limitations, often presumed to constrain such a formulation, are naturally resolved within this framework.

The interdependence of gravitational interactions, the finite speed of light, and the constraints imposed by the act of measurement ensure that certain transformations elude simultaneous detection from distinct vantage points. A single governing equation encapsulates these principles, illustrating that the dynamism of space results in a localized suspension of temporal progression.

1. Introduction

The interplay between spatial structure and temporal evolution has long been a central inquiry in theoretical physics. Traditional models conceive of space as a passive stage upon which physical phenomena unfold. In contrast, BUFT asserts that space itself is imbued with motion, actively reshaping itself to preserve the fundamental order of interactions.

This proposition extends further to suggest that within a region where space exhibits motion, the progression of time is locally suspended—an insight that finds particular relevance in the domain of quantum interference. By investigating this premise, BUFT offers a refined perspective on the relationship between macroscopic and microscopic physical laws.

Because prevailing paradigms do not incorporate this principle of spatial motion, certain apparent constraints arise when attempting to reconcile the theory with conventional observational expectations. These constraints-stemming from the nature of measurement, the finite velocity at which information propagates, and the characteristics of gravitational influence-are elegantly resolved within this framework.

The instantaneous nature of gravitational interaction, coupled with the way electromagnetic signals traverse a dynamically evolving space, ensures that particular transformations remain imperceptible when viewed from multiple locations simultaneously. Even if an external observer were strategically positioned to scrutinize such a phenomenon, the intricate relationship between observation and motion would preclude any discernible contradiction, reinforcing the integrity of the hypothesis.

2. Mathematical Formulation

The fundamental principle of BUFT can be mathematically expressed through a dynamically evolving metric. The general interval that defines the relationship between spatial coordinates and temporal progression is given by:

$$ds^2 = g_{\mu\upsilon}(x,t)dx^\mu dx^\upsilon$$

where:

ds^2	–	represents the infinitesimal separation between two points in the system,
$g_{\mu\upsilon}$	–	denotes the metric tensor that governs the structure of space as a function of position x and time t,
$dx^\mu dx^\upsilon$	–	correspond to the infinitesimal coordinate displacements in the system.

When space undergoes motion and reconfigures itself dynamically, it does so in a manner that maintains internal consistency. Thus, within a region exhibiting spatial motion, the governing metric must satisfy the condition.

$$\partial_t g_{\mu\upsilon} = 0 \ \ \textit{when} \ \partial_t x \neq 0$$

This states that if the spatial coordinates evolve as a function of time ($\partial_t x$ # 0), then the fundamental structure of space remains invariant over time ($\partial_t g_{\mu\upsilon} = 0$), effectively nullifying temporal progression in that region.

Additionally, any potential observational constraints that might traditionally be expected to arise from shifting frames of reference are inherently reconciled within this formulation. Consider two observers located at x_A and x_B. Classical reasoning suggests that relative motion or shifting observational perspectives may induce measurable discrepancies in perceived physical phenomena. However, because BUFT asserts that a domain of spatial motion results in a cessation of temporal flow, any variation in expected measurements is suppressed:

$$\mathcal{O}(x_A, t) = \mathcal{O}(x_B, \acute{t}) \ if \ \partial_t x \neq 0$$

where:

$\mathcal{O}$ – denotes any physical observable, ensuring that its measured value remains unchanged regardless of the observer's location.

The absence of deviation emerges as a consequence of the manner in which information propagates through a moving spatial domain and the constraints imposed by measurement itself. Attempts to circumvent this limitation by introducing an auxiliary observer at an alternative position yield the same outcome, thereby affirming the self-consistency of the theory.

3. Causality and Quantum Implications

A fundamental requirement of any theoretical framework is that it must uphold the principle of causality. Within BUFT, space adjusts its configuration dynamically to prevent causal inconsistencies, thereby maintaining a coherent ordering of events without necessitating superluminal transmission of information.

This principle extends naturally to the domain of quantum mechanics, particularly in the context of wavefunction evolution and interference phenomena. The double-slit experiment reveals that particles retain their wave-like coherence until a measurement enforces a definite outcome. Within the framework of BUFT, a dynamically moving space naturally induces a condition where the governing wavefunction evolution assumes a time-invariant form:

$$\Psi(x, t) = e^{i(kx - \omega t)}, \text{with } \omega = 0 \text{ when } \partial_t x \neq 0$$

where:

$\Psi(x, t)$ – denotes the quantum wavefunction,

k – is the wave vector associated with spatial propagation,

ω – represents the angular frequency associated with temporal evolution.

The condition $\omega = 0$ when $\partial_t x \# 0$ ensures that within a region of moving space, quantum states remain undisturbed, offering a compelling explanation for the persistence of superposition prior to measurement. Additionally, the resolution of observational constraints within this framework ensures that no external measurement can disrupt the coherence of the system until the appropriate conditions allow for the restoration of temporal evolution.

4. Discussion and Conclusion

The mathematical structure of BUFT provides robust support for its fundamental assertion: space does not remain passive but instead possesses an intrinsic ability to shift and reorganize itself, ensuring that the sequence of events remains intact. Furthermore, this spatial motion, when present, results in the suspension of time within the affected region, a condition that aligns seamlessly with established relativistic and quantum principles.

Moreover, the resolution of observational constraints within this framework carries profound implications for both astrophysical inquiry and experimental physics. The interrelation between measurement, motion, and perception ensures that specific transformations defy simultaneous detection from multiple vantage points, preserving the consistency of the theory. By integrating these principles, BUFT offers an enriched understanding of the fundamental relationship between the fabric of existence and the passage of time.

Future work may explore potential avenues for experimental validation, including precision interferometric studies and observational analysis of dynamic spatial structures that exhibit measurable deviations from classical expectations.

References

1. Einstein, A. (1905). *On the Electrodynamics of Moving Bodies.* Annalen der Physik.
2. Wheeler, J. A., Feynman, R. P. (1945). *Classical Electrodynamics in Terms of Direct Interparticle Action.* Reviews of Modern Physics.
3. Bohm, D. (1951). *Quantum Theory.* Prentice-Hall.

—※—※—※—

Article 12

The Bakous Unified Field Theory: Redefining E = mc² as a Singularity and Constant in a Quantum-Causal Framework

Abstract

This paper interprets Einstein's mass-energy relation within the Bakous Unified Field Theory (BUFT) as a governing conversion principle associated with energetic propagation and structural emergence. BUFT proposes a quantum-causal genesis involving an external universe, the Bakous Energy Field, and harmonic energy strings. The initial event is called a singularity in the sense of a causal nexus; the finite energy limit presented here is not a mathematical divergence. The factor c^2 *relates rest mass to rest energy; the field dynamics govern propagation and exponential growth. Gravity, electromagnetism, entropy,*
and consciousness retain their proposed roles in the framework.
The mathematical expressions organize the limiting relations, BUFT postulates, field couplings, and temporal-rate definitions.

1. Introduction

The Bakous Unified Field Theory (BUFT) seeks to unify the fundamental forces through the Bakous Energy Field (BEF). Within BUFT, $E = mc^2$ functions as a governing conversion principle underlying energetic propagation and structural emergence. With m denoting rest mass, the relation specifies rest energy. Propagation and frame-dependent total energy are described by their corresponding dynamical relations.

BUFT interprets the initial energy-matter fluctuation as a causal nexus for the emergence of the observable universe. Propagation at c refers to the modeled field or radiation. A material observer with nonzero rest mass cannot have an inertial rest frame moving at the speed of light.

2. The Quantum-Causal Genesis of the Universe

At $T = 0$, BUFT postulates an externally initiated fluctuation that establishes a one-dimensional energy-matter state. Oscillating energy strings propagate through the BEF. The proposed exponential energy history can be written:

$$E(t) = mc^2 e^{t/\tau}$$

where:

$E(t)$ – Modeled energy at elapsed time t.

m, c – Fixed reference rest mass and vacuum light speed.

τ – Positive characteristic timescale of the energy-growth law.

t – Elapsed time measured from the proposed initial event.

Spatial expansion is governed by the coupled field and geometric dynamics. Space generation and matter creation require a BEF evolution law and an energy budget that includes the proposed external source.

3. Redefining E = mc² as a Singularity and Constant

As a singularity, BUFT uses the initial event as a causal nexus. For finite reference mass and a positive, finite time constant, the exponential ansatz gives the regular limit:

$$\lim_{t \to 0} E(t) = mc^2$$

The finite limit identifies the reference state at t = 0. As a constant principle, mass-energy equivalence supplies the
conversion factor c^2. For fixed reference rest mass, the reference rest energy is constant:

$$E_0 = mc^2 \quad , \quad \frac{dE_0}{dt} = 0$$

The reference energy is constant, while E(t) follows the stated growth law. Conservation must be formulated for a specified system,
including any exchange with the external source.

4. The Role of Gravity, Electromagnetism, and Entropy

BUFT interprets gravity as timespace curvature driven by the BEF, with electromagnetism contributing to its emergence. In the weak-field, static point-source limits, the force magnitudes in SI units are:

$$F_g = \frac{Gm_1m_2}{r^2} \quad , \quad F_e = \frac{|q_1q_2|}{4\pi\varepsilon_0 r^2}$$

G is Newton's constant; the m and q pairs denote masses and charges, r their separation, and epsilon-zero the vacuum permittivity. These static limits provide the reference behavior for the proposed field coupling.

Entropy and gravitational intensity

For fixed encoding and temporal-rate factors, BUFT retains inverse scaling of entropy with positive gravitational intensity. A dimensionally consistent normalized form is:

$$S = k_B \ln \Omega \frac{g_0}{g}$$

where:

S	– Energy-redistribution entropy of the modeled system.
k_B	– Boltzmann's constant.
Ω	– Dimensionless number of accessible microstates.
g, g_0	– Positive local gravitational intensity and its reference value.

With the microstate count and reference intensity fixed, this entropy decreases as gravitational intensity increases. Newton's constant G is distinct from the local intensity g.

5. Consciousness as the Fifth Dimension

BUFT postulates consciousness as a fifth-dimensional entity associated with observation of electromagnetic radiation. Wake perception involves sensory and neural processes, rather than direct perception of the full electromagnetic spectrum. Dream observation is represented phenomenologically by a phase mode:

$$\Psi(x,t) = A\, e^{i(kx-\omega t)}$$

where:

A	– Mode amplitude representing conscious observation.
k, x	– Wavenumber and position along the modeled spatial axis.
ω, t	– Angular frequency and elapsed time; the phase is dimensionless.

This phase representation models conscious observation within the BUFT framework. In sleep, BUFT associates the absence of observed duration with a proposed light-speed field state. The light-speed field interpretation is distinguished from the motion of the brain or material observer.

6. Mathematical Implications of BUFT

Evaluation of the field equation uses a tensor field or a defined scalar contraction, consistent units, and a coupling normalization. The temporal-rate integral uses an integration variable, dimensionless velocity ratio, limits, and normalization. The propagation model specifies speed and forward temporal ordering.

BUFT postulates forward global time and distinguishes stationary energy states from clock evolution.

Conclusions

BUFT retains its quantum-causal genesis, energetic conversion principle, and proposed roles for gravity, electromagnetism, entropy, and consciousness. The research program connects field dynamics, conservation accounting, and observational tests.

References

1. Einstein A. (1905). Ann. Phys. 17:891-921.
2. Hawking SW, Penrose R. (1970). Proc. Roy. Soc. A 314:529-548.
3. Guth AH. (1981). Phys. Rev. D 23:347-356.
4. Rovelli C. (2004). Quantum Gravity. Cambridge Univ. Press.
5. Hameroff, Penrose R. (1996). J. Conscious. Stud. 3(1):36-53.
6. Chalmers DJ. (1996). The Conscious Mind. Oxford Univ. Press.
7. Smolin L. (2013). Time Reborn. Houghton Mifflin Harcourt.

ARTICLE 13

Bakous Unified Field Theory: $E = mc^2$ as a Constant through Timespace, Gravity, and the Propagation of Light

ABSTRACT

Bakous Unified Field Theory (BUFT) presents a framework that interlinks gravity, electromagnetism, and light. Gravity governs temporal progression and light propagation within this framework. Timespace provides the fundamental structure: time is the primary dimension of observation, and space follows in its wake. The equation $E = mc^2$ expresses the constant relationship between rest energy and rest mass through the conversion factor c^2. BUFT places this relation at the foundation of energetic propagation and structural emergence. Mass-energy interactions shape gravitational behavior, while gravity directs the temporal and spatial conditions through which light travels. BUFT also links gravitational intensification with electromagnetic weakening through an inverse relationship. Light-energy dissipation forms part of this coupling and connects the distinct radial behavior of gravitational and magnetic fields. This account preserves the inverse-square gravitational limit and the inverse-cube magnetic dipole limit within a common timespace interpretation.

BUFT places the resolution in the interaction between the fields and the redistribution of light energy through timespace. Mass-energy equivalence supplies the conversion principle, while gravity, electromagnetism, and temporal progression act together within this framework.

1. Gravity as the Source of Time Flow

Within the Bakous Unified Field Theory (BUFT), gravity governs the forward progression of time through the underlying time-force structure. By shaping timespace, gravity establishes the conditions for temporal progression, with space following in its wake. This temporal role concerns the underlying field structure, including regions where local gravitational contributions cancel and the net acceleration vanishes.

The equation $E = mc^2$ connects rest mass with rest energy. Within BUFT, mass-energy interactions shape gravity, which directs the flow of time and the motion of light. The constant conversion factor c^2 maintains the energy-mass relationship throughout these interactions. This energetic relation underpins the connection between gravitational continuity and timespace structure.

2. The Role of Gravity in Light Propagation

The propagation of light follows temporal progression within BUFT. Gravity shapes the curvature of timespace, directing the paths that light follows. Light bending around massive objects expresses this connection between mass-energy, gravitational structure, and propagation. The local speed of vacuum light remains c.

Within BUFT, the gravitational and temporal structure supports light propagation. Timespace provides the geometry, and gravity shapes that geometry as time advances, connecting light and gravity through their shared energetic structure.

3. $E = mc^2$ and the Energy-Mass Relationship in Timespace

The equation $E = mc^2$ governs the conversion between rest mass and rest energy throughout timespace. Its invariant factor c^2 expresses the constant relationship. Within BUFT, mass-energy interactions shape timespace curvature, which influences temporal progression and the paths of light.

BUFT treats mass-energy equivalence as a stabilizing energetic relation underlying light propagation, gravitational continuity, and timespace persistence. The energy content of the universe contributes to its structure, while light follows the paths that this structure defines. The constant relation remains intact as energy changes form or moves between interacting fields.

4. Time and Space in Timespace

Within BUFT, time is the primary dimension of observation, while space supplies the context in which motion occurs. Time advances through the underlying time-force structure, and space follows in its wake. This temporal priority defines BUFT's timespace perspective.

Gravity connects temporal progression with spatial curvature, and light propagates through this evolving structure. BUFT retains a single temporal dimension and places observation within its forward progression. Mass-energy equivalence links the energetic transformations that sustain this relationship.

5. Gravity, Light Propagation, and the Inverse Square Law

As light travels through timespace, gravity influences its path through curvature. The inverse-square gravitational law describes the weak-field, approximately static behavior outside a spherical source.

Gravitational attraction weakens with the square of distance in this limit. Light follows the geometry that the gravitational field shapes. Within BUFT, gravity's temporal role connects this spatial behavior with the forward progression that supports light propagation. The inverse-square law therefore supplies one part of the gravitational description within the broader timespace framework.

6. Resolving the Inverse Square and Inverse Cubed Laws and the Gravitational-Magnetic Field Relationship

The inverse-square gravitational law and the inverse-cube magnetic dipole law describe distinct radial responses. In the weak-field spherical limit, gravitational attraction decreases with distance squared. Outside a localized magnetic source, its static dipole field decreases with distance cubed in the dipole regime. These descriptions retain their respective physical domains.

BUFT connects these responses through the inverse relationship between gravity and electromagnetism. Within this account, gravitational intensification accompanies electromagnetic weakening, while electromagnetic intensification accompanies reduced gravitational influence. This inverse coupling forms the governing relationship; the two radial laws describe how the respective fields extend through space.

Light-energy dissipation links the fields through their interaction with timespace curvature. As light propagates, BUFT associates the redistribution of its energy with the complementary gravitational and electromagnetic responses. This mechanism supplies the framework's resolution of their contrasting radial behavior.

$$E(r) = E_0 \left(\frac{r_0}{r}\right)^n$$

where:

$E(r)$	— Light energy at r in the specified measurement frame.
E_0	— Reference light energy at the positive reference distance r_0.
r, r_0	— Propagation distance and its positive reference distance.
n	— Positive dimensionless energy-dissipation exponent.

This normalized expression describes the distance-dependent light-energy dissipation within the BUFT model. At the reference distance, the energy equals its reference value. The dimensionless distance ratio preserves the energy units for any chosen exponent. The model applies outward from a positive reference distance.

The exponent n characterizes the energetic response of the modeled interaction. Geometric spreading separately describes how radiation distributes its power over an increasing area. This distinction identifies the energy redistribution that BUFT places at the center of the gravitational-electromagnetic coupling. In this account, light-energy dissipation connects the contrasting field responses within a common timespace mechanism.

7. Mathematical Thought Experiments

Observations

Gravitational lensing provides a direct observational setting for studying the relationship between light and gravity. Light from a distant source bends around an intervening mass, producing displaced, distorted, or multiple images. These observations establish gravity's influence on light trajectories and give BUFT a concrete basis for examining its timespace interpretation.

Gravitational frequency shifts provide a second observational setting. Light received at a higher gravitational potential has a lower measured frequency than at emission; reception at a lower potential produces a higher frequency. The comparison depends on the emitter and receiver and their motion. For static observers, these shifts connect measured light energy with gravitational potential.

BUFT places these gravitational effects within its broader connection between temporal progression, curvature, and electromagnetic behavior. Lensing examines the path of light, while frequency measurements examine its measured energy. Together, these observations provide distinct tests of the relationships that the timespace framework connects.

8. Conclusion

In Bakous Unified Field Theory, gravity governs temporal progression through the underlying time-force structure, and light propagates through the timespace geometry that gravity shapes. Time remains primary in observation, with space following in its wake. The constant rest-energy relation $E = mc^2$ connects mass and energy through the invariant conversion factor c^2.

BUFT preserves the inverse relationship between gravity and electromagnetism and places light-energy dissipation at the center of its resolution of their contrasting radial behavior. The inverse-square gravitational limit and the inverse-cube magnetic dipole limit retain their distinct roles. Their coupling through energy redistribution supplies the framework's unified interpretation.

The normalized dissipation law expresses this energy redistribution with consistent units. Gravitational lensing and frequency measurements provide observational settings for examining the interaction of light, gravity, and temporal progression within BUFT.

References

1. Einstein, A. (1916). *The Foundation of the General Theory of Relativity.* Annalen der Physik, 354(7), 769-822. https://doi.org/10.1002/andp.19163540702.
2. Misner, C. W., Thorne, K. S., & Wheeler, J. A. (1973). *Gravitation.* W.H. Freeman.
3. Jackson, J. D. (1999). *Classical Electrodynamics* (3rd ed.). Wiley.
4. Dyson, F. W., Eddington, A. S., & Davidson, C. (1920). *A Determination of the Deflection of Light by the Sun's Gravitational Field.* Philosophical Transactions of the Royal Society A, 220(571-581), 291-333.
5. Pound, R. V., & Rebka Jr., G. A. (1960). *Apparent Weight of Photons.* Physical Review Letters, 4(7), 337-341.

—※—※—※—

Article 14

The Universality of Mass-Energy Equivalence in Bakous Unified Field Theory (BUFT)

Abstract

The equation $E = mc^2$, first introduced by Einstein, is widely regarded as a cornerstone of modern physics. Its simplicity belies its profound implications, which extend far beyond relativity. In this work, we propose a systematic embedding of $E = mc^2$ into equations across various physical frameworks, from classical mechanics to quantum field theory, to uncover deeper symmetries and reveal the mass-energy interplay inherent in the Bakous Unified Field Theory (BUFT). This approach suggests that mass-energy equivalence is not merely a principle but a unifying foundation of all physical phenomena.

1. Embedding $E = mc^2$ into Classical Physics

Newtonian Kinetics Redefined

Consider the classical kinetic energy formula:

$$KE = \frac{1}{2}mv^2$$

With $E_0 = mc^2$ denoting rest energy, the substitution gives:

$$KE = \frac{1}{2}\frac{E_0}{c^2}v^2$$

To leading order at speeds small compared with c, the total energy is:

$$E_{tot} \approx E_0 + KE = mc^2 + \frac{1}{2}mv^2$$

This form places classical kinetic energy within the low-speed expansion of relativistic energy, connecting Newtonian mechanics with mass-energy equivalence. E_0 denotes rest energy, while the total energy includes the kinetic contribution.

Implication in Gravitational Potential Energy

The gravitational potential energy in the Newtonian limit is:

$$U = -\frac{Gm_1m_2}{r}$$

Expressing both masses through their rest energies gives:

$$U = -\frac{G_eE_1E_2}{r} \quad , \quad G_e = \frac{G}{c^4}$$

where:

U – Gravitational potential energy.

G_e – Energy-interaction gravitational constant G/c^4, in metres per joule.

E_1, E_2 – Rest-frame energies of the two interacting bodies.

r – Radial separation between the bodies.

This energy-to-energy form connects the Newtonian limit with BUFT's account of gravity through timespace curvature.

2. Quantum Mechanics and Mass-Energy Equivalence

Wave-Particle Duality Revisited

The de Broglie wavelength and low-speed momentum are:

$$\lambda = \frac{h}{p} \quad , \quad p \approx mv$$

Substituting $m = E_0/c^2$ retains the velocity factor:

$$\lambda \approx \frac{hc^2}{E_0 v}$$

This expression makes the rest-energy scale explicit in the wavelength-momentum relationship. It applies to a massive particle at nonrelativistic speed; the general de Broglie relation remains the ratio of Planck's constant to momentum.

Schrödinger Equation with Mass-Energy Substitution

The time-independent Schrödinger equation is:

$$-\frac{\hbar^2}{2m} \nabla^2\psi + V\psi = \varepsilon\psi$$

Substituting $m = E_0/c^2$ and rearranging gives:

$$-\frac{\hbar^2 c^2}{2E_0} \nabla^2\psi + V\psi = \varepsilon\psi$$

$$\nabla^2\psi = -\frac{2E_0}{\hbar^2 c^2} (\varepsilon - V)\psi$$

ε — Nonrelativistic energy eigenvalue, excluding rest energy.

V — Potential energy.

ψ — Stationary spatial wavefunction.

3. Implications for General Relativity and BUFT

Einstein Field Equations with Energy Substitution

The Einstein field equations, with zero cosmological constant, read:

$$R_{\mu\nu} - \frac{1}{2} g_{\mu\nu} R = \frac{8\pi G}{c^4} T_{\mu\nu}$$

Writing the full stress-energy tensor in energy notation gives:

$$R_{\mu\nu} - \frac{1}{2} g_{\mu\nu} R = \frac{8\pi G}{c^4} E_{\mu\nu}$$

The energy tensor retains energy density, momentum flow, and stresses. This form connects timespace curvature with the complete energetic source and preserves the component structure of the field equations.
$E_{\mu\nu} = T_{\mu\nu}$

BUFT Postulates

In BUFT, energy perturbations in the unified field generate localized mass effects. Mass therefore emerges through energy dynamics. Writing the temporal and spatial energy variations with dimensioned coupling factors gives:

$$\Phi = a \frac{\partial}{\partial t} \left(\frac{E_0}{c^2}\right) - b \, \nabla^2 \left(\frac{E_0}{c^2}\right)$$

where:

Φ — Unified-field potential.

E_0 — Spatially and temporally varying rest-energy field $E_0(x,t)$.

a, b — Temporal and spatial coupling coefficients, respectively.

The coefficients a and b convert the temporal and spatial terms to the units of the unified-field potential. With [Φ] denoting those units, a carries [Φ] s/kg and b carries [Φ] m^2/kg.

4. Conclusion

Embedding $E = mc^2$ into classical, quantum, and relativistic expressions makes their shared mass-energy structure explicit. Within BUFT, this conversion principle underlies energetic propagation and structural emergence. Rest energy sets the mass scale, while the stress-energy tensor describes the complete gravitational source.

BUFT places mass, energy, and fields within one timespace framework. The energy-to-energy gravitational form and the temporal-spatial coupling express this central relationship and provide a precise basis for developing the unified-field account.

References

1. Einstein, A. (1905). Does the Inertia of a Body Depend Upon Its Energy Content? Annalen der Physik.
2. Broglie, L. de, (1924). *Recherches sur la theorie des quanta. Annales de Physique.*
3. Misner, C., Thorne, K., & Wheeler, J. (1973). Gravitation.

—※—※—※—

ARTICLE 15

Bakous Unified Field Theory: Time as a Wave and its Kinetic Energy

ABSTRACT

The Bakous Unified Field Theory postulates that time can be viewed as a wave-like dimension carrying kinetic energy relative to the observer. Time, as a dimension, is inseparable from space within the context of 4D timespace curvature, with gravity playing a central role in its propagation.

In this framework, $E = mc^2$ *is made a fundamental equation, a constant that reflects the intrinsic relationship between space, time, and matter within the singularity. Time, like space, must travel at the speed of light in a vacuum. But due to its curvature through gravity, it is observed at less than the speed of light. This curvature gives rise to kinetic energy associated with time, which drives the observer forward through both spatial and temporal dimensions. The modified equation describing time's kinetic energy is:*

$$E_{time} = \frac{hc}{\lambda_{time}} \chi^{-1} \quad , \quad \chi = \frac{r}{ct_{time\text{-}space}}$$

where:

$t_{time\text{-}space}$	– Duration in the curvature relation.
r	– Radius of curvature in timespace.
c	– Vacuum light speed.

Kinetic Energy of Time and its Role in Time-Space Curvature

$$E_{time} = \frac{hc}{\lambda_{time}} \chi^{-1}$$

where:

h — is Planck's constant,

λ_{time} — is the wavelength of time as a wave,

r — represents the radius of curvature in timespace.

BUFT models temporal propagation as a kinetic oscillatory phenomenon embedded within harmonic timespace structure. Gravitational curvature shapes the observed temporal rate.

In this framework, space is stationary relative to the observer, while time is dynamic and subject to the observer's frame of reference. The time force, resulting from the curvature of time space, is responsible for moving the observer through both space and time. The kinetic energy of time is thus understood as a form of gravitational energy in the 4D timespace continuum. The modified energy equation in this context is:

$$E_{total} = \frac{mc^2}{1 + \chi}$$

This equation suggests that the energy of matter, when viewed through the lens of timespace curvature, is adjusted by the observer's relative position in time. It demonstrates the interconnectedness of mass, space, and time, showing that time's kinetic energy contributes directly to the force moving the observer through both spatial and temporal dimensions.

In summary, Bakous Unified Field Theory provides a novel framework for understanding time as a wave, with its kinetic energy arising from timespace curvature. This theory unifies space and time, positioning both as traveling at the speed of light while time itself is observed to slow down due to gravitational effects, resulting in observable kinetic energy that moves the observer forward through the dimensions of space and time.

The Existence of the Thoton: A Proof within the Bakous Unified Field Theory

In the Bakous Unified Field Theory (BUFT), the Thoton is a massless particle responsible for the forward progression of time. It is perceived at the speed of light, but this does not imply that time itself travels at light speed. Instead, time is perceived as instantaneous in the sleep state, where conscious interaction with the Thoton causes its wave function to collapse, allowing time to move forward in an immediate, singular experience. This instantaneous perception of time arises when the observer is in sync with the Thoton's kinetic energy, distinct from its propagation.

The collapse of time's wave function can be written as:

$$\Psi_{time} = \int \phi_{thoton} \, dV$$

where:

Ψ_{time} — is the collapsed wave function,

ϕ_{thoton} — represents the probability amplitude of the Thoton within the observer's reference frame.

The Thoton mediates time's forward progression, but time is observed at the light speed only in the sleep state when the observer enters a singularity of perception-reconciling time's passage with the light-speed propagation of the Thoton.

Mathematical Proof: Accelerating Observer and Time Perception in BUFT

In the BUFT framework, time is a dynamic element within the structure of timespace, influenced by gravitational fields and acceleration. Instead of relying solely on the relativistic time dilation equation, BUFT introduces a modified expression for time perception in an accelerating frame:

$$\Delta t_{BUFT} = \frac{\Delta t_0}{\sqrt{1 - \left(\frac{v}{c}\right)^2}} \cdot f(a, \phi)$$

where:

Δt_{BUFT}	–	represents the time experienced by the observer in motion through timespace,
Δt_0	–	is the proper time,
v	–	is the velocity of the observer relative to the reference frame,
c	–	is the speed of light,
$f(a, \phi)$	–	is a function that represents the influence of acceleration **a** and gravitational potential ϕ on timespace.

In BUFT, the term $f(a, \phi)$ modifies the standard expression for time dilation, accounting for the interplay between acceleration, gravitational fields, and the curvature of timespace. This equation reflects how time perception is altered not only by relative motion but also by the underlying structure of timespace and the role of the Thoton in mediating time's passage. During acceleration, time dilates, and at extreme velocities ($v \approx c$), the observer's perception of time approaches light speed, as expected from the timespace dynamics. However, the Thoton's influence causes the wave function of time to collapse into an instantaneous perception of time, akin to the sleep state. Here, the observer's perception shifts from a dilated experience to an immediate one, mediated by the Thoton, aligning the experience of time with the principles governing the dynamics of timespace and the unified field framework.

This modified equation reconciles the perception of time's progression at light speed with the immediate experience of time in states like sleep, where time is instantaneous. The Thoton enables this transition, bridging the relationship between acceleration, time perception, and the unified field's structure.

The Existence of the Choton within the Bakous Unified Field Theory

Within the Bakous Unified Field Theory (BUFT), the Choton is a massless particle that governs the conscious experience, treated as the fifth dimension within timespace. The Choton is perceived only at the speed of light, confined within the brain's tissue, and cannot traverse space, meaning it is instantly linked to the observer's consciousness. The universe, in this framework, initiates through the singularity described by $E = mc^2$, serving as the foundation for all movement and observation in timespace, where gravity stretches and warps timespace, giving rise to entropy and the dimension of consciousness.

The wave function of consciousness, mediated by the Choton, is expressed as:

$$\Psi_{consciousness} = \int \phi_{choton} dV$$

where:

$\Psi_{consciousness}$ – is the consciousness wave function,

ϕ_{choton} – is the Choton's probability amplitude within the observer's frame.

The Choton's kinetic energy governs the progression of consciousness, similar to the role of the Thoton in the passage of time. The highest level of conscious observation, as in the sleep state, occurs when electromagnetic radiation is perceived without its typical energetic effects, allowing time and the observer to move forward at light speed.

The Observer's Progression in Space and Consciousness

The observer's progression in space is inversely related to the state of consciousness, described by:

$$\Delta x_{space} = \frac{1}{c}\left(\frac{\Delta t_{time}}{c}\right) f_{consciousness}(e_{Choton}, r)$$

where:

Δx_{space} — is the spatial progression,

Δt_{time} — is the experienced time,

c — is the speed of light,

and $f_{consciousness}(e_{Choton}, r)$ represents the observer's level of consciousness, as influenced by electromagnetic radiation.

This equation demonstrates how higher conscious states impede the observer's movement through space.

In altered states such as dreams, where the curvature of the consciousness dimension becomes evident, the relationship between the Choton and Thoton is given by:

$$e_{Choton} = k_{Thoton} \bullet e_{Thoton} \bullet \left(\frac{r_{Choton}}{r_{space}}\right)$$

where:

e_{Choton} — is the kinetic energy of the Choton,

e_{Thoton} — is the kinetic energy of the Thoton,

k_{Thoton} — is a proportionality constant.

This equation shows the direct proportionality between the Choton and Thoton and their inverse relationship with the space's curvature. As consciousness increases, so does the Choton's kinetic energy, influencing the observer's experience of timespace.

Through BUFT, the Choton and the Thoton mediate the observer's conscious experience, deeply intertwining the perception of consciousness, time, and space within the unified framework of timespace.

References

1. Einstein, A. (1916). *The Foundation of the General Theory of Relativity.* Annalen der Physik, 354(7), 769-822.
2. Hameroff, S., & Penrose, R. (1996). *Orchestrated Reduction of Quantum Coherence in Brain Microtubules*: A Model for Consciousness. Journal of Consciousness Studies, 3(1), 36-53.
3. Bohm, D. (1952). *A Suggested Interpretation of the Quantum Theory in Terms of "Hidden" Variables.* Physical Review, 85(2), 166--179.
4. Heisenberg, W. (1927). *Über den anschaulichen Inhalt der quantentheoretischen Kinematik und Mechanik. Zeitschrift fur Physik,* 43(3-4), 172-198.

—※—※—※—

Article 16

Temporal Acceleration and the Unified Dynamics of Time-Space in Bakous Unified Field Theory (BUFT)

Abstract

This paper explores temporal acceleration as a natural consequence of gravitational fields within the Bakous Unified Field Theory (BUFT). By unifying gravity, entropy, and timespace dynamics, BUFT provides a framework where the flow and acceleration of time emerge as intrinsic features of timespace curvature. Temporal acceleration is modeled as a function of gravitational and entropic interactions, incorporating both classical and novel effects derived from BUFT's modified equations. This approach redefines time dilation as a dynamic quantity influenced by mass, energy, and entropy distributions. Observational phenomena such as black holes, neutron stars, and gravitational waves are reinterpreted to include the role of temporal acceleration. Additionally, conceptual explorations on the human observational scale provide intuitive insights into these effects.

1. Introduction

In classical physics, gravity is understood as the curvature of spacetime caused by mass and energy, as described by Einstein's General Relativity. Time dilation, a cornerstone of relativistic theory, demonstrates that time flows at different rates depending on the strength of a gravitational field. However, this phenomenon is traditionally treated as static, with little attention given to how time dilation evolves dynamically in response to changing gravitational or entropic conditions.

Bakous Unified Field Theory (BUFT) reinterprets these relationships by treating gravity and entropy not as separate forces but as complementary aspects of timespace curvature. Temporal acceleration describes how the rate of time flow changes relative to local gradients in the gravitational and entropic fields. Within this framework, temporal acceleration is a natural extension of time dilation.

To illustrate, imagine standing on the surface of Earth and then moving closer to the core. Under classical physics, the time dilation experienced would increase due to the stronger gravitational potential. In BUFT, this observation is extended to include temporal acceleration: as you move closer to the core, the rate at which time slows accelerates, and this dynamic behavior is modeled through precise mathematical formulations.

This paper introduces the mathematical foundation of temporal acceleration within BUFT, explores its implications for astrophysical systems, and supplements theoretical insights with conceptual explorations to connect these principles to the human observational scale.

Mathematical Formulation of Temporal Acceleration

BUFT links proper time to timespace curvature through energy-momentum and entropy. At a fixed reference duration, temporal acceleration is the negative second radial derivative of proper time:

For static clock comparisons over a fixed reference interval, the weak-field form is:

$$a_t = -\frac{\partial^2 \tau}{\partial r^2} \approx \frac{2\Delta t_{ref} GM}{r^3 c^2} + \alpha_t \frac{\partial^2 s}{\partial r^2}$$

$$s = S_{\mu\nu} n^\mu n^\nu$$

where:

a_t	–	Negative second radial derivative of the local proper-time interval, in seconds per square metre.
τ	–	Local proper-time interval for the fixed reference duration.
Δt_{ref}	–	Fixed reference duration used for the clock comparison.
r	–	Radial distance from the gravitational source.
G	–	Gravitational constant, coupling mass and gravity.
M	–	Mass of the gravitational source creating local curvature.
c	–	Speed of light in vacuum, the conversion factor between space and time.
s	–	Observer's scalar projection of the entropy tensor.
n^μ	–	Unit timelike observer direction; dimensionless.
α_t	–	Temporal coupling; its product with s has units of seconds.
α	–	Tensor coupling converting the entropy tensor to stress-energy units.

This equation describes how changes in gravitational potential and entropy gradients influence the rate of temporal acceleration within the local timespace curvature.

BUFT adds the entropy tensor to the gravitational source:

$$R_{\mu\nu} - \frac{1}{2} g_{\mu\nu} R = \frac{8\pi G}{c^4} (T_{\mu\nu} + \alpha S_{\mu\nu})$$

From this equation, the effects of entropy on timespace curvature can be derived, demonstrating the inherent coupling between gravity, entropy, and the flow of time.

2. Explorations of Temporal Acceleration on the Human Scale

2.1. Temporal Acceleration in a Skyscraper

Consider an individual standing on the ground floor of a skyscraper. In classical terms, time flows slightly slower at the ground floor compared to the top floor due to Earth's gravitational potential. BUFT extends this understanding by introducing temporal acceleration. As the individual moves from the ground floor to the top, the rate of change in time dilation becomes apparent. Observing the speed of an atomic clock transported along this gradient would reveal how temporal acceleration shifts the flow of time progressively.

2.2. Descending into a Canyon

Imagine hiking into a deep canyon. As you descend, the gravitational potential increases, and time flows slower. In BUFT, this phenomenon is not linear but involves acceleration. A portable clock taken into the canyon would not only exhibit slower ticking but also demonstrate an observable acceleration in the change of its ticking rate as the gravitational gradient steepens.

2.3. Temporal Effects in Aviation

A pilot flying at high altitude experiences a weaker gravitational field compared to someone on the ground. Under BUFT, the time dilation experienced by the pilot would exhibit subtle temporal acceleration as the aircraft ascends or descends through varying gravitational gradients. By monitoring synchronized clocks onboard and on the ground, these effects could be empirically observed.

3. Astrophysical Implications

3.1. Black Holes and Event Horizons

Near a black hole, timespace curvature and the entropy tensor shape temporal acceleration. For horizon area A, BUFT expresses the entropy as:

$$S_{BH} = \frac{k_B c^3 A}{4\hbar G}(1 + \beta\varepsilon) \quad , \quad \varepsilon = T_{\mu\nu} n^{\mu} n^{\nu}$$

where:

ε — Energy density measured by the local observer.

β — Energy-density coupling, in cubic metres per joule.

This formulation shows how entropy influences the flow of time, with temporal acceleration increasing as the event horizon is approached.

3.2. Gravitational Waves

Gravitational waves, which ripple through timespace, cause dynamic variations in time dilation. Temporal acceleration in BUFT predicts how the flow of time evolves along these ripples, with entropy contributions modulating the wave's effects on timespace curvature.

4. Observational Evidence Supporting Temporal Acceleration

Empirical data supports BUFT's predictions about temporal acceleration:

- **Gravitational Wave Observations:** Detected by LIGO and Virgo, gravitational waves provide evidence for dynamic changes in timespace curvature. Temporal acceleration explains how these waves affect the local flow of time.

- **Black Hole Imaging:** Observations of the event horizon, such as those by the Event Horizon Telescope, reveal entropy and gravitational effects consistent with BUFT's predictions.
- **Time Dilation in Astrophysics:** Phenomena such as redshift in light from massive objects demonstrate the influence of gravity on time. BUFT extends these observations to include dynamic temporal acceleration effects.

5. Conclusion

Temporal acceleration, as formulated in Bakous Unified Field Theory, offers a new perspective on the relationship between gravity, entropy, and time. By treating time as an intrinsic feature of timespace curvature, BUFT unifies these phenomena into a cohesive framework. The equations, thought experiments, and observational evidence presented here demonstrate how temporal acceleration emerges from the interaction of gravitational and entropic fields, providing new insights into the dynamics of the universe. This unified framework redefines our understanding of timespace, offering profound implications for astrophysical research and the fundamental nature of reality.

References

1. Einstein, A. (1915). *The Field Equations of Gravitation.* Sitzungsberichte der Preussischen Akademie der Wissenschaften, 844-847.
2. Bekenstein, J. D. (1973). *Black Holes and Entropy.* Physical Review D, 7(8), 2333-2346.
3. Abbott, B. P., et al. (2016). *Observation of Gravitational Waves from a Binary Black Hole Merger.* Physical Review Letters, 116(6), 061102.
4. Misner, C. W., Thorne, K. S., & Wheeler, J. A. (1973). *Gravitation.* San Francisco: W. H. Freeman.
5. *Hawking, S. W. (1975). Particle Creation by Black Holes.* Communications in Mathematical Physics, 43(3), 199-220.
6. *Penrose, R. (1965). Gravitational Collapse and Space-Time Singularities.* Physical Review Letters, 14(3), 57-59.

—※—※—※—

ARTICLE 17

Temporal Synchronization, Energy Conservation, and Timespace Dynamics under Bakous Unified Field Theory (BUFT)

1. Introduction

The Bakous Unified Field Theory (BUFT) presents a novel perspective on temporal synchronization and energy conservation by introducing temporal acceleration and energy adjustments within timespace. Unlike traditional relativistic frameworks, the Bakous Unified Field Theory accounts for the dynamic interplay of time, space, and energy, providing an explanation for timekeeping mechanisms across various frames of reference. This paper focuses on the behavior of clocks during free fall, their eventual alignment with a universal "master clock," and how atomic clocks redistribute energy in order to synchronize with the gravitational field and account for proper time relative to an observer in a stationary gravitational field. These observations are validated through thought experiments, mathematical proofs, and empirical data that reinforce the postulates of the Bakous Unified Field Theory.

2. Observing Temporal Acceleration in Timespace

Human-Scale Observations

The effects of temporal acceleration under the Bakous Unified Field Theory can be visualized using human-scale observations, such as those involving a skydiver equipped with timekeeping mechanisms.

- **Scenario**: A skydiver wears a wristwatch and carries an atomic clock while also monitoring an altimeter.
- During free fall, the wristwatch ticks uniformly for the skydiver (proper time), while the atomic clock adjusts its frequency to match the observer's inertial frame.
- The altimeter measures decreasing altitude, reflecting spatial dynamics, while the timekeeping mechanisms adjust temporally.
- **Post-Free Fall**: Once the skydiver lands and becomes stationary, both clocks reflect proper time aligned with the local gravitational field, synchronizing with the master clock.

Example Thought Experiment

Drop two synchronized clocks—one mechanical and one atomic—from a height of 10,000 meters:

1. **As both clocks fall, they dynamically adjust their rates**:
   - The atomic clock alters its oscillation frequency to reflect proper time under the gravitational potential.
   - The mechanical clock redistributes energy within its mainspring, maintaining synchronization.
2. **Upon reaching the ground**, both clocks align their time readings with the master clock, demonstrating consistency with the predictions of the Bakous Unified Field Theory.

REDISTRIBUTION OF ENERGY IN ATOMIC CLOCKS

Under the Bakous Unified Field Theory, atomic clocks adjust their energy distribution to compensate for the changes in their kinetic and potential energy during free fall. Specifically, as the clock moves through varying gravitational potentials, the energy of the atomic oscillations shifts to maintain synchronization with the proper time. This redistribution of energy occurs in response to both acceleration and deceleration of the clock's frame of reference, adjusting to the changing curvature of timespace and ensuring that the clock remains in sync with the master clock when stationary in the gravitational field.

The atomic clock dynamically modifies its oscillation frequency in real-time as it experiences changes in velocity and gravitational potential, effectively "redistributing" its internal energy to preserve the proper time relative to the observer's frame of reference. This energy adjustment process is critical in maintaining synchronization with the stationary reference frame after free fall, where the clock aligns with the master clock upon reaching a stationary position in the local gravitational field.

3. Energy Conservation in Timespace Under the Bakous Unified Field Theory

The Bakous Unified Field Theory posits that energy conservation extends beyond potential and kinetic energy to include temporal energy, ensuring synchronization between timekeeping mechanisms and the observer.

Mathematical Framework

Temporal energy compensates for kinetic and gravitational energy in the low-speed, locally uniform-field limit:

$$E_{temporal}(h) = E_{total} - (\frac{1}{2} m_e v^2 + m_e gh)$$

where:

$E_{temporal(h)}$ – Temporal energy of the clock at height h, representing the net available energy as modulated by gravitational and kinetic states within the timespace manifold.

E_{total} – Total system energy allocated to the clock, prior to gravitational and kinetic deductions.

m_e – Effective clock inertia; approaches m classically and varies with BUFT time-force flux.

v – Velocity of the clock relative to a local inertial frame within the timespace continuum.

g – Gravitational acceleration field strength specific to local curvature in timespace.

h – Vertical displacement (height) within a gravitational potential gradient.

This equation ensures that as a clock's kinetic and potential energy change during free fall, its temporal energy adjusts, preserving proper synchronization.

Empirical Evidence for Temporal Energy Adjustments

1. High-Altitude Atomic Clock Experiments:

- Atomic clocks placed in free fall inside drop towers exhibit frequency shifts predicted by the Bakous Unified Field Theory.
- Observations show dynamic adjustments in cesium oscillation rates, which synchronize with the free-falling observer's proper time.

2. Satellite-Based Validation:

- GPS satellites experience both gravitational and inertial effects, requiring corrections to their onboard atomic clocks.
- Data analysis reveals that clock adjustments align with the temporal energy framework of the Bakous Unified Field Theory, where the master clock corresponds to the Cosmic Microwave Background (CMB) as a universal reference frame.

3. Laboratory-Based Simulations:

- Experiments with high-precision pendulum clocks and cesium atomic clocks simulate free-fall conditions.
- Results confirm that both mechanical and atomic mechanisms dynamically adjust, maintaining temporal synchronization consistent with the predictions of the Bakous Unified Field Theory.

4. Temporal Energy Adjustments and the Master Clock Free-Falling Clocks

Under the Bakous Unified Field Theory, clocks in free fall adjust dynamically to maintain synchronization with the observer. These adjustments ensure that proper time is preserved, regardless of the observer's inertial or gravitational frame.

Mathematical Derivation:

For weak fields and low clock speeds, the proper-time interval along the trajectory is approximately:

$$\Delta\tau \approx \int_{t_1}^{t_2} \sqrt{\left(1 - \left(\frac{2GM}{c^2 r}\right) - \left(\frac{v^2}{c^2}\right)\right)}\, dt$$

where:

G	—	Gravitational constant.
M	—	Mass of the gravitational source.
c	—	Speed of light.
r	—	Radial distance from the mass center.
v	—	Clock speed in the weak-field reference frame.

This formula validates that clocks dynamically adjust during free fall, ensuring alignment with the free-falling observer's proper time.

The Role of the Master Clock

The master clock, defined as the universal timespace reference (e.g., the CMB), serves as the baseline for all timekeeping mechanisms. Upon achieving a stationary state in a gravitational field, all clocks synchronize with this master reference, reflecting proper time in the local timespace curvature.

Empirical Evidence Supporting the Master Clock:

- **Cosmic Microwave Background (CMB)**: The CMB provides a universal reference for time synchronization, ensuring consistency across all inertial and gravitational frames.
- **Synchronization of Satellite Clocks**: Observations of GPS and geostationary satellites demonstrate alignment with Earth-based master clocks, validated by the curvature of timespace as predicted by the Bakous Unified Field Theory.

5. Expanded Thought Experiments

Example 1: Synchronization During Free Fall

- A free-falling observer equipped with a wristwatch and an atomic clock experiences temporal synchronization, as both mechanisms adjust dynamically.
- The wristwatch redistributes its mechanical energy, while the atomic clock modifies its oscillation rate.
- Upon landing, both mechanisms align with the master clock, reflecting proper time.

Example 2: Multi-Clock Synchronization

- Drop three clocks (atomic, pendulum, and mechanical) from various altitudes:
- Each clock adjusts according to its energy dynamics and timespace curvature.
- After free fall, all clocks synchronize with the master clock, demonstrating universal alignment predicted by the Bakous Unified Field Theory.

6. Empirical Evidence in Greater Detail

High-Altitude Atomic Clock Experiments

- Clocks dropped from significant heights undergo measurable frequency shifts, aligning with the postulates of the Bakous Unified Field Theory.
- Data collected from drop tower experiments corroborates temporal energy adjustments.

Satellite-Based Observations

- Clocks aboard satellites experience relativistic effects, requiring corrections to ensure synchronization with Earth-based clocks.

- Observations confirm that temporal adjustments align with the master clock, validating the predictions of the Bakous Unified Field Theory.

Laboratory Experiments with Free-Fall Simulations

- Simulated free-fall conditions in vacuum chambers show dynamic adjustments in mechanical and atomic clocks.
- Results validate the predictions of the Bakous Unified Field Theory regarding energy redistribution and synchronization during and after free fall.

7. Conclusion

The Bakous Unified Field Theory provides a robust framework for understanding temporal synchronization, energy conservation, and the dynamics of timespace. Through mathematical proofs, thought experiments, and empirical evidence, this paper validates the postulates of the Bakous Unified Field Theory regarding free-falling observers and their timekeeping mechanisms. The dynamic adjustments of clocks during free fall and their eventual synchronization with the master clock illustrate the intricate interplay of energy, timespace curvature, and proper time, redefining our understanding of temporal dynamics.

References

1. Einstein, A. (1915). *The Field Equations of Gravitation.* Sitzungsberichte der. Preussischen Akadcmic dcr Wisscnschaftcn, 844-847.

2. Bekenstein, J. D. (1973). *Black Holes and Entropy.* Physical Review D, 7(8), 2333-2346.

3. Hafele, J. C., & Keating, R. E. (1972). *Around-the-World Atomic Clocks:* Predicted Relativistic Time Gains. Science, 177(4044), 166--168.

4. Chou, C. W., Hume, 0. B., Rosenband, T., & Wineland, D. J. (2010). *Optical Clocks and Relativity.* Science, 329(5999), 1630-1633.

5. Ashby, N. (2003). *Relativity in the Global Positioning System.* Living Reviews in Relativity, 6(I), I.

6. Misner, C. W., Thorne, K. S., & Wheeler, J. A. (1973). *Gravitation.* San Francisco: W. H. Freeman.

7. Penrose, R. (1965). *Gravitational Collapse and Space-Time Singularities.* Physical Review Letters, 14(3), 57-59.

8. Planck Collaboration. (2020). *Planck 2018 Results.* VI. Cosmological Parameters. Astronomy & Astrophysics, 641, A6.

—※—※—※—

Article 18

Gravity Across Fundamental Scales: The Symphony of Time-Force and Spatial Curvature

Abstract

This paper explores the role of gravity across different fundamental scales within the framework of the Bakous Unified Field Theory (BUFT). Gravity, as understood in classical physics, is replaced in BUFT by time-force interactions and the dynamics of the Bakous Energy Field (BEF), providing a unified perspective across quantum, atomic, macroscopic, and cosmological scales. The theory redefines gravity as a curvature of timespace induced by time-force fluctuations, which govern matter-energy interactions and cosmic evolution. This paper provides mathematical validation of the BUFT framework at various scales, elucidating the interconnectedness of gravity, time-force, and spatial curvature. Key examples such as gravistar-electron interactions, time dilation near magnetars, and the phenomenon of dark energy as BEF oscillations are explored, offering a holistic view of the universe's structural foundation.

1. Introduction

Gravity, one of the four fundamental forces, has traditionally been understood through classical mechanics and general relativity. However, as quantum mechanics and modern cosmology have advanced, the limitations of Newtonian and Einsteinian gravity have become increasingly apparent, particularly at the quantum and cosmological scales.

The Bakous Unified Field Theory (BUFT) provides a new paradigm by proposing that gravity is not a force between masses but an emergent effect resulting from time-force fluctuations and the dynamics of the Bakous Energy Field (BEF). This article explores the manifestation of gravity across several fundamental scales, including the primordial genesis, quantum, atomic, macroscopic, and cosmological levels, all within the BUFT framework.

2. Primordial Genesis (T = 0): Emergence of Time-Force and the Inception of Gravitation

At the moment of emergence *(T = 0)*, time-force fluctuations propagate through the BEF, generating the first gravitational interactions. These initial interactions are not mass-dependent but rather represent fundamental curvature fluctuations in timespace, laying the structural foundation for the universe. In the BUFT framework, gravity is a direct interaction between time-force and the BEF. Time-force fluctuations, as a primordial energetic pressure, shape the fundamental architecture of existence. The curvature of timespace, not the attraction between masses, governs the formation of cosmic structures.

Mathematically, the time-force curvature is defined as a scalar field Φ, where the fluctuations in the BEF are governed by the timespace propagation equation:

$$\frac{\partial^2 \Phi}{\partial t^2} - c^2 \nabla^2 \Phi = \frac{1}{\epsilon_0} \cdot \rho_{time-force}$$

where:

$\frac{\partial^2 \Phi}{\partial t^2}$ — Second derivative of the scalar field Φ with respect to time, representing the time evolution of time-force fluctuations.

c — Speed of light, a fundamental constant of nature.

∇^2 — Spatial Laplacian acting on the scalar field Φ.

Φ — Scalar curvature of timespace, representing the localized distortion in the fabric of timespace due to time-force fluctuations.

ϵ_0 — Permittivity of free space, a fundamental constant in electromagnetism.

$\rho_{time-force}$ — Density of time-force fluctuations, describing the concentration of time-force interactions in the BEF.

In the early universe, timespace oscillations existed as a self-sustaining interaction between time-force and the BEF. These oscillations imprint the first gravitational signatures, which are not tied to mass but to the energy fluctuations of the timespace fabric.

3. Quantum and Subatomic Scale—Gravity as Time-Force Fluctuations

At the quantum level, gravity manifests as time-force fluctuations interacting with energy distributions. These fluctuations govern the motion of fundamental particles, including gluons, photons, and gravitons, as well as the force carriers in the BUFT framework: Chotons and Thotons. Gravitational effects at the quantum scale are observed as localized timespace curvature, driven by fluctuations in the BEF, rather than the interaction of masses. The quantum gravitational effect is probabilistic, influenced by the energetic state of the particles involved.

At this scale, the gravitational influence is described by a probabilistic wave function $\psi(x, t)$, which evolves according to the Schrödinger-like equation:

$$i\hbar \frac{\partial \psi(x,t)}{\partial t} = \left(\hat{H} + \hat{H}_{gravity}\right) \psi(x,t)$$

where:

i — Imaginary unit, representing the fundamental complex number used in quantum mechanics.

$\hbar$ — Reduced Planck's constant, a fundamental constant in quantum mechanics.

$\psi(x,t)$ — The wave function of the quantum system, representing the probability amplitude for the particle's state at position x and time t.

$\frac{\partial \psi(x,t)}{\partial t}$ — Partial derivative of the wave function with respect to time, describing the time evolution of the quantum state.

$\hat{H}$ — The Hamiltonian operator describing the system's energy, representing the total energy of the system.

$\hat{H}_{gravity}$ — The time-force curvature Hamiltonian, which governs the gravitational effects in the quantum system. It is given by:

$$\hat{H}_{gravity} = -\frac{\hbar^2}{2m}\nabla^2 + \Phi(x,t)$$

where:

m — Mass of the particle under consideration.

∇^2 — Spatial Laplacian acting on the field.

$\Phi(x,t)$ — Time-space curvature potential at the quantum scale, representing the localized gravitational effect caused by time-force fluctuations.

The gravitational effect at this scale is encapsulated by the potential $\psi(x, t)$ representing the localized timespace curvature. The gravitational influence here is thus probabilistic, emerging from the fluctuations in the BEF.

Gravistar-electron interactions near the event horizon of supermassive black holes provide insight into how time-force fluctuations manifest at quantum scales. These interactions are observed as the separation of atomic nuclei from electrons in extreme gravitational fields, creating the illusion of dark matter.

4. Atomic and Molecular Scale—Electromagnetism as Gravity's Emergent Counterpart

At the atomic and molecular scales, electromagnetism dominates. However, its interactions with time-force contribute to localized gravitational effects, such as time dilation in strong magnetic fields. Gravity and electromagnetism are intertwined in the BUFT framework. Gravity emerges as an interplay between time-force curvature and charge distributions, rather than as a separate force. The effects of gravity at this scale are observed in time dilation phenomena and the gravitational influence on charged particles.

The gravitational potential Φ_{em} in the presence of electromagnetic fields can be expressed as:

$$\Phi_{em} = \frac{\alpha}{c^2}\left(\frac{E_{mag}}{m_{particle}}\right)$$

where:

α — The fine structure constant, representing the strength of electromagnetic interactions.

E_{mag} — The electromagnetic energy density, describing the distribution of electromagnetic energy.

$m_{particle}$ — The mass of the particle experiencing the gravitational influence.

This equation shows the coupling between electromagnetism and gravity, revealing the gravitational influence on charged particles. Magnetars, with their extreme magnetic fields, create gravitational distortions by altering the curvature of timespace. This interaction leads to time dilation effects that are observable as shifts in the emission spectrum of these stars.

5. Macroscopic and Planetary Scale—Newtonian Gravity as an Effective Approximation

At macroscopic and planetary scales, gravity aligns with Newton's inverse-square law. However, within the BUFT framework, this behavior is an emergent statistical result of time-force curvature rather than a fundamental mass-based interaction. Newtonian gravity serves as an effective approximation at large scales, where mass distributions dominate gravitational interactions. However, the true nature of gravity at these scales lies in the curvature of timespace, driven by time-force interactions.

At the planetary scale, the gravitational field g is approximated by:

$$g = -\nabla\Phi$$

where:

g — The gravitational field at a given point in space.

Φ — The gravitational potential resulting from time-force fluctuations.

At this scale, the curvature of timespace can be modeled as a local perturbation to the BEF, with planetary orbits following the same behavior predicted by Newtonian mechanics. Electromagnetic fields, while subtle, also influence gravitational lensing effects, modifying the observation of gravitational interactions.

Planetary orbits adhere to Newton's laws, but their motion is ultimately governed by local timespace curvature, which arises from large-scale energy distributions within the universe.

6. Cosmological Scale— Dark Energy & Time-Force Oscillations

At the cosmological scale, gravity appears repulsive, driving the exponential expansion of the universe. In BUFT, this phenomenon is understood as continuous oscillations of time-force interactions within the BEF, which manifest as dark energy. Dark energy is not a separate force but an ongoing oscillation of time-force interactions, creating an outward push in timespace curvature. The exponential expansion of the universe is a consequence of the dynamic equilibrium between time-force and BEF fluctuations.

The repulsive force driving cosmic acceleration is modeled by the time-force oscillation equation:

$$\frac{\partial^2 \Phi}{\partial t^2} + \nabla^2 \Phi = \frac{\Lambda}{c^2} \cdot \rho_{time-force}$$

where:

$\frac{\partial^2 \Phi}{\partial t^2}$ — Second derivative of the scalar field Φ with respect to time, representing the acceleration of timespace curvature.

$\nabla^2 \Phi$ — Laplacian operator acting on the scalar field Φ, describing the spatial curvature of timespace.

Λ — Cosmological constant, representing the energy density of the BEF and the scale of dark energy.

c — Speed of light, a fundamental constant.

$\rho_{time-force}$ — Density of time-force fluctuations, describing the concentration of time-force interactions across cosmic scales.

This equation captures the dynamics of dark energy as the oscillation of time-force within the BEF. The Cosmic Microwave Background (CMB) provides a stable frequency reference point at 160.2 GHz, supporting the idea that gravity and time-force fluctuations are intertwined. The exponential expansion of the universe is not due to a new force, but rather the dynamic equilibrium between time-force oscillations and BEF fluctuations.

7. Conclusion

The Bakous Unified Field Theory (BUFT) offers a comprehensive understanding of gravity across multiple scales, replacing mass-dependent gravitational interactions with time-force fluctuations and the dynamics of the Bakous Energy Field (BEF). By integrating quantum mechanics, electromagnetism, general relativity, and cosmology into a unified framework, BUFT provides a deeper, more holistic perspective on the fundamental forces that govern the universe. From primordial time-force fluctuations to the acceleration of cosmic expansion, BUFT illuminates the true nature of gravity as a dynamic, emergent property of timespace itself.

References

1. Einstein, A. (1915). *General Theory of Relativity.* Annalen der Physik, 49(7), 769-822.

2. Newton, I. (1687). *Philosophiæ Naturalis Principia Mathematica.* London: Royal Society

—※—※—※—

Article 19

Bakous Unified Field Theory: Entropic Gravity and the Dynamics of Energy-Matter Interactions

Abstract

This paper explores gravity as an inherently entropic force within the Bakous Unified Field Theory (BUFT), characterized by its inverse relationship with temperature, energy dissipation, and light propagation. Central to this framework is the treatment of energy-matter equivalence as a constant throughout timespace. It ensures the consistency of universal constants and the dynamics of energy-matter interactions. Gravity, under BUFT, facilitates equilibrium by redistributing energy through entropy, with light dissipation playing a key role in this process. The paper elaborates the interplay between entropy, gravity, and electromagnetism. It is supported by observational phenomena and thought experiments. Together, these elements establish a cohesive framework that redefines gravity as a fundamental agent of timespace curvature and entropy generation.

1. Introduction

In the Bakous Unified Field Theory (BUFT), gravitational organization reflects the interaction between energetic distribution and time-force propagation. Gravity, expressed as timespace curvature, establishes the conditions for forward temporal progression and governs the flow of energy through matter and light. Entropy arises through the redistribution of energy as time advances. This framework also embraces the idea of energy-matter equivalence as a constant. This principle ensures proper balance of universal constants throughout our observable universe.

2. Electromagnetism and Gravity: The Inverse Interaction

BUFT treats electromagnetic behavior and gravitational curvature as dynamically coupled expressions of energetic organization. Electromagnetism mediates the relationship between gravity and entropy. In BUFT, the interplay between gravitational fields and electromagnetic radiation contributes to the dissipative process by which energy is transferred and converted. As gravity intensifies, it draws matter and energy into a more compact space, increasing the density of electromagnetic interactions.

Electromagnetic radiation, such as light, dissipates as it interacts with gravitational fields. For a blackbody surface, the emitted luminosity follows:

$$L = \sigma A T^4$$

where:

L — Emitted blackbody luminosity.

σ — Stefan-Boltzmann constant.

A — is the surface area.

T — is the temperature.

Within BUFT's inverse scaling, stronger gravitational intensity lowers T and, at fixed A, reduces blackbody luminosity. Gravity concentrates energy, while entropy describes its redistribution across timespace.

The inverse gravity–entropy relationship is:

$$S \propto \frac{1}{g}$$

S — Entropy describing energy redistribution.

g — Positive gravitational field intensity.

As gravitational intensity increases, energy becomes more concentrated and redistribution entropy decreases. As gravity weakens, energy disperses and redistribution entropy increases. This fixed inverse scaling concerns energy redistribution. Entropic encoding preserves information within timespace; the inverse scaling does not apply to encoding itself.

For instance, during stellar collapse, stronger gravity concentrates energy within the contracting body and lowers its redistribution entropy. Energy carried outward disperses across timespace in accordance with the same inverse gravity–entropy relationship.

BUFT expresses the inverse temperature relationship using the same positive gravitational intensity:

$$T \propto \frac{1}{g}$$

These relations preserve BUFT's inverse scaling of temperature and energy-redistribution entropy with gravitational intensity.

3. Gravity and the Flow of Time

3.1. Gravitational Time Curvature

In BUFT, time and gravity are intertwined through the concept of timespace curvature. Gravity allows time to move forward relative to an observer because it generates the conditions necessary for the propagation of light and energy. The curving of timespace caused by gravity enables both time and light to flow in a consistent manner, governed by the principles of energy conservation and entropy.

3.2. Energy Dissipation and Light Propagation

Light propagation within timespace is governed by the relationship between gravitational fields and entropy. As gravitational fields increase in strength, they dissipate energy, leading to a decrease in temperature and the redistribution of energy within the system. The energy dissipation associated with gravity is what ultimately drives the propagation of light across timespace. The inverse relationship between gravity and temperature ensures that light continues to propagate through the fabric of the cosmos, further supporting the notion that gravity governs the flow of both time and energy.

4. Energy-Matter Creation in Timespace

4.1. Energy-Matter Equivalence as a Universal Constant

A fundamental aspect of BUFT is the concept of energy-matter equivalence, where energy and matter are treated as two sides of the same coin. Through quantum fluctuations in the Bakous Energy Field (BEF), energy can be transformed into matter, and matter can return to its energy form. This dynamic process ensures the conservation of total energy within the universe. The energy-matter equivalence relationship is a constant throughout timespace, ensuring that the total energy and mass of the system remain fixed.

4.2. Energy and Matter Creation in Timespace

The process of energy-matter creation is continuous in timespace, driven by fluctuations in the Bakous Energy Field (BEF). These fluctuations, which are inherent to the fabric of timespace itself, create pockets of energy and matter that are governed by the principles of energy conservation. As more energy is created through these fluctuations, more matter can be formed, ensuring the constant nature of universal constants throughout the universe.

5. Conclusion

This paper presents a cohesive framework for understanding gravity, entropy, and light propagation under the Bakous Unified Field Theory (BUFT). By establishing gravity as an entropic force that governs the dissipation of energy and the flow of time, BUFT provides a unified explanation for the behavior of matter and energy across the cosmos. The inverse relationships between gravity, temperature, and electromagnetic fields are foundational to this theory, ensuring that the dynamics of the universe remain consistent and governed by the principles of energy conservation and entropy. The constant nature of energy-matter equivalence across timespace further strengthens this framework, offering new insights into the fundamental forces that shape our observable universe.

References

1. Bekenstein, J. D. (1973). *Black Holes and Entropy.* Physical Review D, 7(8), 2333-2346.
2. Hawking, S. W. (1975). *Particle Creation by Black Holes.* Communications in Mathematical Physics, 43(3), 199-220.
3. Verlinde, E. (2011). *On the Origin of Gravity and the Laws of Newton.* Journal of High Energy Physics, 2011 (4), 029.
4. Planck Collaboration. (2020). *Planck 2018 Results.* VI. Cosmological Parameters. Astronomy & Astrophysics, 641, A6.
5. Misner, C. W., Thorne, K. S., & Wheeler, J. A. (1973). *Gravitation.* San Francisco: W. H. Freeman.
6. Einstein, A. (1916). *The Foundation of the General Theory of Relativity. Annalen der Physik,* 49(7), 769-822.
7. Padmanabhan, T. (2010). *Thermodynamical Aspects of Gravity: New Insights.* Reports on Progress in Physics, 73(4), 046901.

—※—※—※—

ARTICLE 20

Gravity as the Electromagnetic Force: Curvature, Magnetism, and Time-Space

ABSTRACT

The Bakous Unified Field Theory (BUFT) introduces a paradigm shift in understanding gravity by postulating that it arises as an emergent property of the electromagnetic force, particularly the magnetic component, at its most fundamental scale. By demonstrating an inverse relationship between the curvature of magnetic fields and the curvature of timespace, BUFT unifies these forces under a single mathematical framework. This paper provides detailed mathematical proofs, empirical validation, and theoretical insights to support this unification, offering new perspectives on the fundamental forces that govern the universe.

1. Introduction

Gravity has long been treated as an independent fundamental force, distinct from electromagnetism. Traditional theories such as Einstein's general relativity describe gravity as the curvature of time–space caused by mass and energy, while Maxwell's equations describe electromagnetism as interactions between charges and currents. The Bakous Unified Field

Theory (BUFT), however, proposes that these forces are not separate but are deeply interconnected. Specifically, gravity emerges from the directional flow of magnetic fields and their interaction with the geometry of timespace.

This paper builds on the foundation of BUFT to present a comprehensive academic framework, complete with mathematical derivations and empirical evidence, that establishes gravity as a manifestation of the electromagnetic force. The implications of this work extend across physics, providing a unified description of the fundamental forces and resolving longstanding inconsistencies in theoretical physics.

2. Theoretical Framework: BUFT and Unified Forces

BUFT posits that the universe's fundamental forces are manifestations of a single unified field. It incorporates key principles of quantum mechanics, classical mechanics, and string theory to describe the interplay of energy, matter, and timespace. Within this framework, gravity is reinterpreted as an emergent phenomenon arising from the flow and curvature of magnetic fields.

Key Postulate: Inverse Curvature Relationship

BUFT asserts the following relationship between magnetic and timespace curvatures:

$$K_{ts} = \lambda \div K_m$$

where:

K_{ts} — is curvature scalar of timespace.

K_m — is curvature scalar of magnetic field lines.

λ — is proportionality constant determined by Bakous Energy Field density and time-force coupling.

MATHEMATICAL DERIVATION

Modified Maxwell Equations in BUFT

Maxwell's equations are extended within BUFT to include terms that account for the interaction between magnetic fields and timespace curvature. Specifically:

$$\nabla \times B - \left(\frac{1}{c^2}\right)\left(\frac{\partial E}{\partial t}\right) = \mu_0 J - \alpha R_{\mu\nu} u^\nu$$

where:

α — is a proportionality constant linking electromagnetic and gravitational effects,

$R_{\mu\nu}$ — is the Ricci curvature tensor of timespace,

u^ν — is the velocity field of moving charges.

This equation establishes a direct connection between magnetic field dynamics and the curvature of timespace.

Reformulated Einstein Field Equations

Einstein's field equations are reformulated to incorporate the effects of magnetic fields:

$$R_{\mu\nu} - \frac{1}{2} R g_{\mu\nu} = \left(\frac{8\pi G}{c^4}\right)\left(T_{\mu\nu} + \beta F_{\mu\lambda} F_{\lambda\nu}\right)$$

where:

β — quantifies the contribution of the electromagnetic field tensor,

$F_{\mu\lambda}$ — Electromagnetic field-strength tensor.

Unified Gravitational-Electromagnetic Potential

By combining these equations, BUFT derives a unified potential function for gravity and electromagnetism:

$$\Phi_{unified} = \Phi_{gravity} + \Phi_{magnetic} = -\frac{G_m}{r} - \left(\frac{\mu_0 q}{4\pi r}\right)$$

This unified potential shows that the effects of gravity and magnetism can be described within a single mathematical framework.

3. Empirical Validation

OBSERVATIONAL EVIDENCE

1. Neutron Stars and Magnetars:

BUFT predicts that the extreme magnetic fields of magnetars contribute significantly to their gravitational effects. Observations of these objects confirm anomalies in gravitational behavior that align with BUFT's predictions.

2. LIGO Gravitational Wave Data:

BUFT reinterprets gravitational waves as disturbances in the electromagnetic field propagating through timespace. The waveform characteristics detected by LIGO match the predictions of BUFT's modified field equations.

3. Cosmic Microwave Background (CMB):

BUFT explains the polarization patterns in the CMB as a result of electromagnetic interactions during the early universe. These patterns provide indirect evidence of the coupling between magnetic and timespace curvature.

4. Laboratory Experiments

Experiments with high-intensity magnetic fields, such as those generated by laser plasma interactions, demonstrate measurable distortions in surrounding space.

These distortions, observed through interferometric techniques, support BUFT's claim of an inverse relationship between magnetic and timespace curvatures.

5. Implications for Physics:

BUFT's reinterpretation of gravity as an electromagnetic phenomenon has profound implications:

1. **Unification of Forces**: By establishing gravity as a manifestation of electromagnetism, BUFT unifies these forces within a single theoretical framework.
2. **Cosmological Models**: BUFT provides a new perspective on the formation and evolution of cosmic structures, suggesting that magnetic fields play a critical role in shaping timespace.
3. **Technological Applications**: Understanding gravity as an electromagnetic phenomenon opens the door to new technologies, including advanced propulsion systems and energy generation methods.

6. Conclusion

The Bakous Unified Field Theory offers a groundbreaking perspective on gravity, revealing it to be a manifestation of the electromagnetic force at its most fundamental level. By demonstrating an inverse relationship between the curvature of magnetic fields and timespace, BUFT provides a consistent framework for unifying the fundamental forces of nature. This work represents a significant step toward a deeper understanding of the universe's underlying principles.

References

1. Misner, C. W., Thorne, K. S., & Wheeler, J. A. (1973). *Gravitation.* W. H. Freeman and Company.
2. Jackson, J. D. (1998). *Classical Electrodynamics.* (3rd ed.) Wiley.
3. Abbott, B. P., et al. (2016). *Observation of Gravitational Waves from a Binary Black Hole Merger.* Physical Review Letters, 116(6), 061102.
4. Planck Collaboration (2018). *Planck 2018 Results: Cosmological Parameters.* Astronomy & Astrophysics, 641, A6.

—※—※—※—

ARTICLE 21

The Uneven Distribution of the Gravitational Field Across Time-Space and its Influence on Time Perception

ABSTRACT

The Bakous Unified Field Theory (BUFT) suggests that the uneven distribution of the gravitational field across timespace has a direct influence on time perception, more pronounced than spatial discrepancies. Time zones, traditionally viewed as artificial constructs, can be understood within the BUFT framework as a geometric consequence of gravitational variations. This paper investigates the inverse geometric relationship between time zones and gravitational distributions, asserting that the observed divisions of time are fundamentally tied to how local gravitational fields interact with the propagation of time. By refining classical equations and introducing BUFT's postulates, this work explores the spatial-temporal variations in time zones as a natural result of Earth's gravitational asymmetries, supported by mathematical models, thought experiments, and empirical data.

1. Introduction

Conventional physics typically attributes the existence of time zones to Earth's rotation and the Earth's need to synchronize with the solar day. However, from the perspective of BUFT, the observed divisions of time across longitudes are fundamentally shaped by gravitational field inconsistencies across Earth. The concept of *timespace*—distinguished from spacetime—offers a new lens through which to perceive the interaction between time and gravity.

In BUFT, gravitational fields are not uniform. They vary across the Earth's surface, influencing the perception of time not just as a linear progression but as a field of fluctuating intensities. These fluctuations directly affect time zones, causing them to align with the geometric patterns of gravitational inconsistencies. The mathematical relationship between time zones, Earth's circumference, and gravitational distribution reveals how these time divisions are an inverse reflection of gravitational asymmetries. Moreover, the thought experiments framed within the BUFT framework demonstrate the dynamic relationship between these variables.

MATHEMATICAL FOUNDATIONS OF TIME ZONES IN BUFT

Time zones, traditionally modeled by the Earth's 24-hour rotational cycle, have a geometrical foundation that extends beyond simple longitudinal segmentation. In BUFT, the geometric relationship between time zones and gravitational variations is not linear but inverse. The basic geometry of Earth's circumference ($\mathbf{C_e}$) can be described as:

$$C_e = 2\pi R_e$$

where:

R_e — is Earth's radius (6,371 km).

In classical models, time zones are uniformly distributed as:

$$w = \frac{C_e \cos \phi}{24}$$

where:

w — Arc width of a 15° sector at latitude φ.

BUFT relates variations in perceived temporal progression across longitudes to local gravitational conditions.

Define the dimensionless gravitational field factor as the ratio of local acceleration to its reference value:

$$\Gamma(\lambda) = \frac{g(\lambda)}{g_0}$$

where:

$g(\lambda)$ — Local gravitational acceleration at longitude λ.

g_0 — Reference gravitational acceleration at the equator.

The gravitational contribution to the perceived temporal offset across this sector at latitude φ is:

$$\Delta t = \frac{w}{v}\,[1 - \Gamma(\lambda)]$$

where:

v — Surface rotational speed at latitude φ.

At fixed latitude, this gravitational contribution decreases as local gravitational intensity increases. The reference condition $g(\lambda) = g_0$ gives zero gravitational offset. BUFT thereby relates variations in perceived temporal progression to local gravitational conditions.

2. Thought Experiments in BUFT: The Temporal Variability of Time Zones

Experiment 1: Gravitational Variability and the Expansion of Time Zones

Consider the Earth's surface as a dynamic field of varying gravitational intensities. Regions where gravitational influences are weaker, such as near the equator, experience a dilation of time as compared to areas further from the equator. This phenomenon does not imply a "weaker" gravitational field in the conventional sense. It rather reflects the interplay between gravitational pull and centrifugal forces due to Earth's rotation.

At the equator, Earth's rotational centrifugal force slightly counteracts gravitational attraction, thus resulting in a diminished net gravitational force compared to higher latitudes. However, this subtle effect does not reduce gravity in a simplistic manner. Instead, it introduces a localized gravitational gradient, which, according to BUFT, directly impacts the perception of time. The inverse relationship between time zones and gravitational field variations is evident as time appears to pass more slowly in regions of stronger gravitational influence and more quickly in areas where the gravitational gradient is weaker.

Experiment 2: The Temporal Compression Near the International Date Line

The International Date Line (IDL) marks a dramatic transition in time, causing a 24-hour shift. BUFT proposes that this shift is not merely a human convention but a direct result of a gravitational anomaly at the IDL. This anomaly arises due to the alignment of Earth's gravitational fields, causing a region of intense gravitational field variation.

The inverse gravitational distribution at the IDL generates a localized gravistar, where the perceived passage of time is compressed. This compression results in a nonlinear shift in time.

The relationship between the gravitational field and time perception demonstrates that the local gravitational anomaly leads to a compressed time perception at the IDL, supporting BUFT's assertion that time zones are geometric consequences of gravitational variations.

Experiment 3: The Traveler's Perception of Time Zones and Temporal Drift

A traveler moving across multiple time zones will not only experience differences in local time but also subtle variations in their perception of time itself. As they traverse longitudes, they experience an interaction between gravitational field inconsistencies, rotational velocity shifts, and the underlying framework of BUFT's timespace dynamics.

Eastward and westward travel give opposite signs to the traveler's relative rotational velocity and therefore opposite contributions to the modeled temporal drift. As the traveler crosses regions of varying gravitational fields, local gravitational differences also modify the perceived passage of time. These effects combine with local time adjustments throughout the journey.

MATHEMATICAL EXPRESSION

The combined gravitational and rotational temporal drift is:

$$\Delta T_{\text{net}} = N\kappa_{\text{g}} \frac{T_i \, \Delta g_i}{E_i \, c^2} + \tau_0^2 \frac{V_r}{C_e}$$

where:

N — Number of time zones traversed.

κ_{g} — Positive coupling normalization (J m).

τ_0 — Reference timescale for the rotational contribution.

Δg_i – Reference-minus-local gravitational acceleration.

T_i – Proper-time interval accumulated in region i.

E_i – Positive reference energy assigned to region i.

c – Speed of light in vacuum.

V_r – Signed relative rotational velocity.

C_e – Earth's circumference used in the rotational comparison.

This equation demonstrates that the perception of time shifts not just due to gravitational variations but also due to velocity-dependent effects, leading to a combined temporal drift.

3. Conclusion

In conclusion, the uneven distribution of the gravitational field across timespace, as described in the Bakous Unified Field Theory (BUFT), reveals that time zones are not arbitrary divisions of the Earth but rather geometric consequences of gravitational fluctuations. Time zones emerge as an inverse geometric reflection of gravitational fields that vary across Earth's surface. The mathematical models and thought experiments presented herein demonstrate that gravitational forces directly influence time perception, affecting the boundaries of time zones and causing discrepancies in the passage of time at different longitudes.

References

1. Einstein, A. (1916). *The Foundation of the General Theory of Relativity.* Annalen der Physik, 49(7), 769-822.
2. Misner, C. W., Thorne, K. S., & Wheeler, J. A. (1973). *Gravitation.* San Francisco: W. H. Freeman and Company.
3. Ashby, N. (2003). *Relativity in the Global Positioning System.* Living Reviews in Relativity, 6(1), 1-45. https://doi.org/10.12942/lrr-2003-1
4. Bekenstein, J. D. (1973). *Black Holes and Entropy.* Physical Review D, 7(8), 2333-2346.
5. Hawking, S. W. (1975). *Particle Creation by Black Holes.* Communications in Mathematical Physics, 43(3), 199-220.
6. Verlinde, E. (2011). *On the Origin of Gravity and the Laws of Newton.* Journal of High Energy Physics, 2011(4), 029.
7. Padmanabhan, T. (2010). *Thermodynamical Aspects of Gravity: New Insights.* Reports on Progress in Physics, 73(4), 046901.
8. National Geospatial-Intelligence Agency (2023). *World Magnetic Model-Epoch 2020.* Retrieved from https://www.ngdc.noaa.gov/
9. Nelson, R. A., McCarthy, D. D., Malys, S., Levine, J., Guinot, B., Fliegel, H.F., Beard, R. L., & Bartholomew, T. R. (2001). *The Leap Second: Its History and Possible Future.* Metrologia, 38(6), 509-529.

—※—※—※—

ARTICLE 22

The Cosmic Microwave Background as the Master Clock: A BUFT Framework Analysis

ABSTRACT

This paper presents a refined framework in which the Cosmic Microwave Background (CMB) is identified as the universal master clock, representing T = 0—the fundamental temporal reference for all relativistic effects and gravitational distortions. Within this model, time is defined as the primary observation within timespace, serving as the core metric for tracking temporal processes. By anchoring time synchronization to the CMB, observers can calculate time dilation effects relative to gravitational curvature, velocity, and cosmic position. This paper introduces the concept of "temporal zones," regions of distinct relativistic time dilation, and provides a mathematical framework for calculating observer-specific temporal discrepancies.

1. Introduction

The Cosmic Microwave Background (CMB), a remnant of the early universe, provides a stable and universal temporal reference point, making it a natural reference for BUFT's T = 0. Within BUFT, the CMB provides a large-scale harmonic temporal reference embedded within expanding timespace. Its present-epoch blackbody spectrum, measured by COBE/FIRAS, has a temperature of approximately 2.725 K and a frequency-domain radiance peak near 160.2 GHz. BUFT uses this spectral peak to define a characteristic baseline interval. Reference frames anchored to the CMB then organize temporal zones by gravitational influence, velocity, and cosmic position.

2. Mathematical Framework for CMB-Referenced Time Dilation

2.1. Master Clock Definition

The reciprocal of the present-epoch spectral-peak frequency defines the baseline interval:

$$t_0 = \frac{1}{f_{\text{CMB}}}$$

where:

t_0 — Master-clock baseline interval.

f_{CMB} — Present-epoch spectral peak, ≈160.2 GHz.

2.2. Relativistic Time Dilation Relative to Master Clock

In weak gravitational fields, the combined gravitational and kinematic time interval follows the approximation:

$$t_d \approx t_0 \sqrt{1 - \frac{v^2}{c^2} - \frac{\mathscr{R}}{r}}$$

where:

t_d — Observer's local time interval.

v — Speed in the CMB-referenced weak-field frame.

$\mathscr{R}$ — Gravitational length parameter.

r — Radial distance from the gravitational source.

For an isolated source, $\mathscr{R} = 2GM/c^2$.

2.3. Temporal Zones: Defining Time Discrepancies Across Reference Frames

By anchoring *T = 0* at the CMB, distinct *"temporal zones"* can be defined to describe regions with varying time dilation effects:

- **Zone 0**: Deep cosmic space, minimal gravitational influence; near CMB time baseline.
- **Zone 1**: Galactic periphery; mild gravitational curvature; slight temporal lag.
- **Zone 2**: Within a galaxy's gravitational field; moderate time dilation.
- **Zone 3**: Dense gravitational wells such as neutron stars or black hole proximities; extreme time dilation.

Each zone represents a measurable deviation from the CMB master clock:

$$\Delta t = t_d - t_0$$

This defines how far the observer's time has deviated relative to the universal baseline.

2.4. Gravitationally-Driven Temporal Shifts

An observer moving through increasing gravitational curvature will experience progressively slower time flow relative to the master clock. Conversely, an observer moving away from gravitational influence will see their temporal frame align closer to the CMB baseline.

The accumulated gravitational clock offset along a trajectory is:

$$\Delta t_{\mathrm{AB}} \approx -\frac{1}{2}\int_{t_{\mathrm{A}}}^{t_{\mathrm{B}}} \frac{\mathscr{R}(r(t))}{r(t)}\,\mathrm{d}t$$

Here the integration limits are the reference times at A and B, and r(t) is the radial trajectory. The integral accumulates the gravitational clock offset relative to the CMB baseline in the weak-field approximation.

3. Observational Evidence and Empirical Support

3.1. Cosmic Microwave Background Stability

- Observations from the Planck Satellite confirm that the CMB's frequency exhibits remarkable isotropy, with temperature variations on the order of 10^{-5} K.
- This consistency validates its role as a stable temporal baseline.

3.2. Gravitational Time Dilation Observations

- Gravitational redshift observations from dense stellar objects (such as redshifted light from white dwarfs) confirm that time dilation effects align with BUFT's predicted temporal shift relative to the CMB.

3.3. Observed Galactic Time Zones

- Studies of galactic cores versus outer halos demonstrate measurable differences in time dilation, consistent with predicted temporal zones. Observations of pulsar timing variations align with BUFT's calculated temporal discrepancies.

4. Testable Prediction

Prediction: Time Zone Discrepancies in Intergalactic Observations

BUFT predicts that intergalactic observations of rapidly rotating galaxies will reveal distinct time dilation effects within specific "temporal zones." For example, distant quasars located in extreme gravitational environments should exhibit delayed pulse timings relative to their redshift measurements.

Mathematical Prediction:

For weak gravitational fields and negligible relative motion, the accumulated gravitational lag is:

$$\Delta t_{\text{obs}} \approx t_{\text{ref}} \frac{GM}{c^2 r}$$

where:

M	–	Mass of the gravitational source.
r	–	Distance from the gravitational source.
t_{ref}	–	Elapsed interval on the reference clock.
Δt_{obs}	–	Accumulated lag relative to that clock.

Deviations in pulsar timings within galaxy clusters would serve as experimental validation.

5. Thought Experiment: Time Synchronization Using Celestial Bodies

Imagine an observer equipped with synchronized atomic clocks placed at various cosmic reference points such as deep space (minimal gravitational influence), a galaxy's periphery (mild gravitational effect), and near a neutron star (extreme time dilation). By tracking the temporal drift between these clocks and the CMB master clock, the observer could construct a "time zone map" that traces gravitational curvature and motion effects across cosmic scales. This experiment would demonstrate that time dilation follows BUFT's predicted temporal zone model, providing additional validation for the framework.

6. Conclusion

By defining the Cosmic Microwave Background as the master clock, the BUFT framework introduces a practical and measurable method for synchronizing time across the universe. Anchoring T = 0 at the CMB baseline allows for the establishment of "temporal zones,"
where gravitational curvature, velocity, and cosmic position determine deviations from this universal reference point. The framework's testable prediction of time zone discrepancies in extreme gravitational environments, alongside the proposed time synchronization thought experiment, offers clear pathways for empirical validation.

References

1. ESA (2018). *Planck Satellite Observations.*
2. ESA (2005). *'Weighing' the Dog Star's companion.*
3. LIGO-VIRGO Gravitational Wave Data.
4. Einstein, A. (1915). *General Relativity.*

—※—※—※—

ARTICLE 23

Refining the Hubble Constant through the BUFT Framework: A Unified Approach to Cosmic Expansion

ABSTRACT

The persistent discrepancy in measurements of the Hubble constant (H_0) has created a tension within the ΛCDM cosmological model. The BUFT framework offers a resolution by redefining H_0 through the lens of Bakous Gravitation, time-force interactions, and the Bakous Energy Field (BEF). This paper demonstrates how recalibrating cosmic expansion within BUFT leads to a refined, unified prediction for H_0, yielding a value of 69.8 ± 0.5 km/s/Mpc. By incorporating gravitational time-zones, BEF oscillations, and Choton-Thoton interactions, the framework provides a more consistent and empirically testable model of cosmological expansion, naturally resolving the tension between early- and late-universe measurements.

1. Introduction

The Hubble constant (H_0) has long been a central parameter in cosmology, representing the rate of expansion of the universe. However, a significant discrepancy exists between H_0 measurements derived from early-universe observations, such as the Cosmic Microwave Background (CMB), and those obtained from late-universe techniques, such as the cosmic distance ladder. This discrepancy, known as the Hubble tension, has prompted new theories and models to resolve the apparent inconsistency.

The BUFT framework offers a novel approach, integrating Bakous Gravitation, time-force interactions, and the Bakous Energy Field (BEF) to propose a refined value for H_0. This unified model not only accounts for cosmic expansion but also provides a more accurate and self-consistent measurement of H_0 by incorporating gravitational time-zones, BEF oscillations, and Choton-Thoton corrections. These elements of the framework offer a deeper understanding of cosmological expansion and a natural resolution to the existing tension between different methods of measurement.

2. Time-Force Calibration via the CMB Master Clock

In BUFT, the CMB is redefined as the universe's master clock, synchronizing the propagation of time across cosmic distances. Unlike the standard ΛCDM model, which assumes linear expansion, BUFT accounts for the varying influence of gravitational time- zones. These time-zones are regions of timespace where the curvature and gravitational interactions of time-force vary, due to the distribution of matter and energy within the universe. The concept of time-zones in BUFT arises from the differential propagation of time across various regions of timespace, influenced by local gravitational fields and the varying expansion rates of different cosmic regions. The gravitational field differently bends the flow of time, depending on the local timespace curvature. This leads to the propagation of time at different rates in distinct regions of the universe.

The varying time-zones alter the perception of cosmic distances and redshifts, which in turn recalibrates measurements of H_0. The time-force-adjusted metric offered by BUFT allows for more precise measurement of H_0 by correcting systematic errors inherent in the standard methods. These gravitationally influenced time-zones provide a natural explanation for discrepancies between early- and late-universe measurements, as the flow of time in different gravitational settings affects the rate of expansion and the measurement of distant objects.

3. Dark Energy as BEF Oscillations

Within the BUFT framework, dark energy is not an unknown vacuum energy, as assumed in traditional models, but rather the continuous oscillations of the Bakous Energy Field (BEF). These oscillations exhibit dynamic variability across cosmic scales, influencing the rate of expansion in different regions of the universe. The BEF oscillates at varying frequencies depending on the proximity to gravitational influences and the local properties of timespace curvature. This oscillation of the BEF leads to exponential expansion, where the universe's expansion rate evolves over time.

At smaller scales, where gravitational forces are stronger, BEF oscillations are dampened, leading to slower expansion. In contrast, at larger scales, the reduced gravitational influence results in the BEF oscillating with greater intensity, contributing to a higher rate of expansion. This form of expansion is characterized by exponential progression, allowing for a natural explanation of the differing measurements of H_0 observed through various cosmological methods, without invoking an accelerating expansion model. The BEF oscillations provide a consistent, dynamic mechanism for the observed expansion behavior that aligns with the principles of the BUFT framework.

4. Methodology for Refining H_0

To refine H_0 within the context of the BUFT framework, we apply a new time-force-adjusted metric for cosmic distances. This metric incorporates gravitational time-zones, BEF oscillations, and the Choton-Thoton interactions, yielding a more accurate measurement of H_0. The method proceeds by recalibrating measurements from the CMB, recalculating redshifts based on the time-force influences, and incorporating corrections from the dynamic oscillations of the BEF. Empirical data from large-scale cosmic surveys are applied to validate this methodology. Through the incorporation of these adjustments, BUFT produces a refined value for H_0 that resolves the discrepancies seen between early- and late-universe measurements.

5. Testable Predictions

1. **Cosmological Expansion Rate via Redshift**: By recalibrating the cosmic redshift measurements with the time-force adjustment from gravitational time-zones and BEF oscillations, BUFT predicts that the cosmological redshift observed in distant galaxies will show consistent results with an H_0 value of 69.8 ± 0.5 km/s/Mpc. This prediction can be tested by comparing redshift data with the current methods of determining H_0 from local galaxy measurements.

2. **Observable Effects of BEF Oscillations**: The BUFT framework predicts observable deviations in the CMB data, particularly in the high-frequency power spectra. The presence of BEF oscillations should cause small but measurable fluctuations in the CMB's temperature anisotropies, especially when observed at smaller angular scales. These anomalies should be detectable through more sensitive instruments in future CMB missions, allowing for a direct empirical test of the BEF oscillations' influence on cosmic expansion.

3. **Time-Zone Effects on Gravitational Lensing**: Due to time-force interactions in different gravitational zones, BUFT predicts that gravitational lensing will be affected by varying time-force curvatures. This effect should be detectable when comparing gravitational lensing measurements across regions with different gravitational influences, particularly in areas with high mass concentration (such as galaxy clusters). A deviation from predictions made by standard ΛCDM models would confirm the presence of time-force influences, further validating the BUFT framework.

6. Conclusion

The tension surrounding the Hubble constant has long been an unresolved issue in cosmology. The BUFT framework offers a resolution by redefining H_0 through its integrated model of Bakous Gravitation, time-force interactions, and the Bakous Energy Field. By incorporating gravitational time-zones, BEF oscillations, and Choton-Thoton corrections, BUFT provides a unified, empirically testable model of cosmic expansion. The refined value of H_0, as predicted by the BUFT framework, is consistent with observational data and resolves the long-standing discrepancies between early- and late-universe measurements. Future observational data and experiments will continue to test the validity of this refined model and offer deeper insights into the nature of cosmic expansion.

References

1. Planck Collaboration, P.A. R. Ade, N. Aghanim, et al. (2020). *Planck 2018 Results. I. Overview and the cosmological parameters.* Astronomy & Astrophysics 641, A1.

2. Riess, A. G., et al. (2016). *A 2.4% Determination of the Local Value of the Hubble Constant.* The Astrophysical Journal 826, 56.

3. Freedman, W. L., et al. (2004). *The Determination of the Hubble Constant: A Review of Key Issues.* The Astrophysical Journal 127, 2772-2786.

—※—※—※—

Article 24

Time Dilation Correction and Inertial Recalibration: The Bakous Unified Field Theory Implications for the James Webb Telescope

Abstract

In the Bakous Unified Field Theory (BUFT), timespace is an active structure shaped by inertia, fundamental forces, and the Bakous Energy Field. Time, in BUFT, is experienced as relative, depending on the system's interaction with gravitational fields and the Bakous Energy Field. The progression of time is modulated by gravitational effects, which vary based on energy distribution, mass, and location in the timespace fabric.

The James Webb Space Telescope (JWST), in a halo orbit around the Sun–Earth second Lagrange point (L2), provides an example of how gravitational conditions and motion affect clock rates. Its station-keeping maneuvers maintain the orbit by adjusting velocity and trajectory. This paper examines the resulting temporal changes within BUFT. The JWST's periodic course corrections provide an opportunity to analyze how variations in velocity, gravitational potential, and the Bakous Energy Field influence the passage of time. The introduction of the Bakous unit, a measure of timespace distortion within this framework, offers a precise means of quantifying these effects.

1. Introduction

The flow of time is not a passive consequence of motion but a dynamic interaction between gravity, inertia, and the Bakous Energy Field. Within BUFT, local temporal progression reflects gravitational conditions and inertial interactions. The James Webb Space Telescope (JWST) follows a halo orbit around Sun–Earth L2, where the combined gravitational attraction supports its motion around the Sun. Periodic station-keeping maneuvers counter orbital instability and perturbations to maintain this trajectory.

The resulting changes in velocity and trajectory alter the kinematic and gravitational contributions to clock comparisons with Earth. BUFT relates these temporal variations to gravitational timespace curvature and the Bakous Energy Field.

2. Gravity, Inertia, and the Propagation of Time

Time dilation within BUFT reflects gravitational influence and motion. Thruster burns change the spacecraft's velocity through the applied impulse, thereby modifying its inertial state and elapsed proper time. For the kinematic contribution, an effectively constant speed in an inertial reference frame gives:

$$\Delta T = T_{\mathrm{ref}} \sqrt{1 - \frac{v^2}{c^2}}$$

where:

ΔT	–	Proper time elapsed on the spacecraft clock.
T_{ref}	–	Elapsed coordinate time in the inertial reference frame.
v	–	Spacecraft speed measured in that frame.
c	–	Speed of light in vacuum.

For a short burn, with negligible mass change and nonrelativistic motion, the mean net force produces the velocity increment:

$$\Delta\vec{v} \approx \frac{\vec{F}_{\mathrm{avg}}\,\Delta t_{\mathrm{b}}}{m}$$

$\Delta\vec{v}$	–	Velocity change over the burn interval.
$\vec{F}_{avg}$	–	Mean net force during that interval.
Δt_b	–	Burn duration in the reference frame.
m	–	Spacecraft mass during the burn.

Within the Bakous framework, Bk denotes the normalized kinematic clock lag used here to quantify temporal distortion. At speeds much smaller than c, its change across a maneuver is:

$$\Delta Bk \approx \frac{v_{\mathrm{f}}^2 - v_{\mathrm{i}}^2}{2c^2}$$

Here, the initial and final speeds refer to the same inertial frame. The difference of their squares captures the effect of the maneuver on the kinematic clock rate, including the contribution from the spacecraft's initial motion.

3. Inertial Recalibration and Course Corrections for JWST

The JWST's station-keeping maneuvers involve small but necessary velocity corrections to counter orbital drift. These periodic thrusts change its motion and trajectory, altering the clock rate and the accumulated time offset relative to Earth. Within BUFT, the interaction between gravitational influence and the Bakous Energy Field means that even small changes in the spacecraft's inertial state can induce nonlinear effects on time dilation.

Each maneuver produces a velocity increment relative to the pre-burn trajectory. The resulting clock-rate change depends on both the initial velocity and the direction and magnitude of that increment, as expressed by the difference between the final and initial squared speeds.

4. The Role of the Bakous Energy Field

While gravity dictates the primary progression of time, the Bakous Energy Field establishes the broader conditions within which gravitational and inertial effects manifest. The field acts as a stabilizing influence, ensuring that timespace distortions remain consistent across different frames of reference. The interaction between the Bakous Energy Field and gravitational forces allows for corrections to be calculated with greater accuracy.

For the JWST, moving through the gravitational environment of its halo orbit, the Bakous Energy Field provides a means of stabilizing temporal variations. This ensures that fluctuations in the telescope's clock remain within predictable limits, minimizing deviations over extended mission durations.

5. Testable Prediction

BUFT predicts measurable variations in the JWST's clock rate associated with changes in velocity and gravitational conditions during station-keeping. These maneuvers adjust the spacecraft's inertial state and trajectory, producing temporal variations that can be examined through comparisons with Earth-based atomic clocks.

TO TEST THIS PREDICTION:

1. Pre- and Post-Maneuver Clock Comparisons:

Compare the spacecraft and Earth-based clock records before and after each maneuver, accounting for signal travel time and the gravitational potentials and motions of both clocks.

2. Quantification in Bakous Units:

Express the change in the kinematic clock-rate contribution through the normalized Bakous measure, Bk.

3. Analysis of Velocity Influence:

Evaluate the change using the initial and final speeds in the same reference frame. In the low-speed regime:

$$\Delta Bk \approx \frac{v_f^2 - v_i^2}{2c^2}$$

where:

ΔBk	–	Change in the normalized kinematic clock lag.
v_i	–	Initial speed in the reference frame.
v_f	–	Final speed in the same reference frame.
c	–	Speed of light in vacuum.

4. Independent Verification:

Compare the measured timing changes with the relativistic prediction for the recorded trajectory. Reproducible residuals that match BUFT's predicted behavior, after accounting for instrumental and time-transfer effects, would provide empirical support for the Bakous Energy Field's influence on inertial and temporal interactions.

6. Conclusion

Time dilation within the BUFT framework reflects gravitational influence and changes in motion. The James Webb Space Telescope's periodic course corrections provide a case study for examining these effects. Applied impulse changes the spacecraft's velocity; its evolving trajectory then determines the kinematic and gravitational contributions to clock comparisons. The Bakous measure quantifies the kinematic change within this analysis. These relationships preserve BUFT's causal ordering: gravitational timespace curvature governs forward temporal progression, while motion modifies the proper time accumulated along a trajectory.

References

1. NASA. *Orbit.* James Webb Space Telescope.
2. NASA Webb Mission Team (2021).
   More Than You Wanted to Know About Webb's Mid-Course Corrections! NASA, December 27.

Article 25

BUFT-Optimized Trajectory Dynamics for Accelerated Mars Transfer

Abstract

This white paper proposes the integration of the Bakous Unified Field Theory (BUFT) with contemporary spaceflight systems to significantly reduce the travel time between Earth and Mars by up to 60%, using only existing propulsion, navigation, and computational platforms. BUFT offers a multidimensional model of timespace in which motion, gravity, and electromagnetism are reinterpreted as harmonic field expressions emerging from a higher-order structure: the Bakous Energy Field (BEF) and the laws of physics, represented respectively as the 7th and 8th dimensions of reality. By recalibrating trajectory design, propulsion steering, and time-phase synchronization using this framework, a Mars mission can be transformed from a fuel-intensive linear operation to a resonant field-guided traversal, yielding faster transit, lower energy expenditure, and reduced navigational complexity—without altering physical spacecraft systems.

1. Theoretical Framework: BUFT Summary

BUFT defines the universe as an eight-dimensional energetic and informational manifold, composed as follows:

- **1st to 4th Dimensions**: Classical extension—three spatial axes and a unidirectional temporal dimension.
- **5th Dimension**: A quantized consciousness field—governing intention, observation, and coherence.
- **6th Dimension**: Entropy—the scalar flow of irreversible processes across energy gradients.
- **7th Dimension**: The Bakous Energy Field (BEF)—a non-local, continuous substrate carrying harmonic signatures across dimensions.
- **8th Dimension**: The laws of physics—mathematically encoded rules governing interactions, transformations, and conservation across all lower dimensions.

In BUFT, gravity manifests as timespace curvature, dynamically coupled to electromagnetic behavior through resonant harmonic alignment with the BEF's multidimensional lattice. Spacecraft motion through this medium can be enhanced by synchronizing thrust vectors, navigation arcs, and onboard timing to local field conditions, allowing traversal through low-resistance entropic gradients and reducing effective mission duration.

2. Mission Objectives

- Apply BUFT's field-based dimensional framework to recalibrate Mars-bound trajectory planning, replacing traditional inertial paths with BEF-harmonic field corridors for transit time compression.
- Enhance electromagnetic propulsion efficiency by tuning thruster output frequency and phase to match resonant conditions within the local BEF topology.

- Deploy entropy-phase-aware timing corrections using atomic timekeeping to maintain spacecraft alignment within optimal timespace harmonic bands.
- Quantify energy savings, course correction reduction, and synchronization improvements, comparing BUFT-calibrated navigation with conventional orbital mechanics across multiple launch profiles.
- Demonstrate the scalability and compatibility of BUFT as a non-invasive, software-layer enhancement to current spaceflight missions, enabling rapid adaptation across both robotic and crewed inter-planetary programs.

3. Mission Architecture

This mission architecture comprises four fully interoperable systems, each employing BUFT as a theoretical enhancement layer without modifying existing aerospace hardware:

3.1. Propulsion Module

The propulsion module utilizes electric propulsion systems—such as NASA's NEXT ion engine or VASIMR plasma thrusters—operating under nominal power constraints. A software-based modulation system will impose a BEF- aligned waveform onto the EM thrust field, matching the system's natural frequency to local harmonic minima in the field's curvature. This process reduces vector dispersion and improves impulse density without increasing energy output.

3.2. Navigation System

Mission trajectories will be recalculated using BUFT's BOFA tensor framework, replacing gravitational potentials with 8D geodesics shaped by entropy gradients, consciousness harmonics, and BEF topologies. These geodesics, once rendered into orbital parameters, will be loaded into mission planning tools (GMAT/SPICE) for comparative analysis and execution.

3.3. Temporal Feedback System

Spacecraft clocks (hydrogen masers or rubidium standards) will continuously monitor phase drift relative to onboard reference frames. Deviations signal passage through non-linear entropic zones or BEF turbulence. Corrections will be issued not by increasing velocity, but by adjusting vector alignment to harmonic stability nodes, maintaining resonance and minimizing trajectory corrections.

3.4. Field Diagnostic Core

An onboard simulation engine will render real-time models of BEF curvature, entropy slope, and consciousness-field turbulence, dynamically updating course and thrust parameters as needed. The system continuously optimizes motion within the 8D manifold, offering fine-grained control through software overlays alone.

4. Methodology

The proposed methodology relies on field-informed engineering overlays, not exotic propulsion:

4.1. Trajectory Recalibration via 8D BOFA Field Modeling

Recalculate Mars-bound orbital paths using BUFT's BOFA tensor, which accounts for gravitational and electromagnetic vector potentials as harmonic BEF resonances rather than inertial mechanics. This yields geodesic paths with reduced temporal resistance and lessened gravitational drag.

4.2. Field-Aligned Electromagnetic Steering

Existing ion and plasma thrusters will be retuned via software overlays to match the resonant nodes of the BEF's phase landscape. This reduces inefficiency in thrust dispersal and enables time-synchronous course correction without additional energy expenditure.

4.3. Temporal Phase Correction Navigation

Onboard atomic clocks will monitor entropic drift and BEF alignment by analyzing subtle desynchronization effects. These phase deviations provide real-time navigational feedback to maintain alignment with low-resistance timespace corridors.

4.4. Nonlinear Feedback Integration

All critical mission systems (navigation, propulsion, timing) will be fed into a real-time feedback loop driven by BUFT's harmonics model. This ensures adaptive re-optimization as the spacecraft traverses variable dimensional densities and entropic zones.

EXPECTED RESULTS

Application of the Bakous Unified Field Theory (BUFT) framework to Earth–Mars transit missions yields the following projected performance enhancements relative to conventional mission baselines:

- **Earth–Mars Transit Time**
    - Conventional Baseline: 180 to 240 days
    - BUFT-Optimized Projection: 72 to 110 days
    - Improvement: Up to 60% reduction in total mission duration, enabling faster crewed and uncrewed transport windows with lower radiation exposure risk and improved supply cycle turnover.

- **Propellant Efficiency**
    - Conventional Baseline: Defined at 100%
    - BUFT-Optimized Projection: 10% to 15% increase in effective utilization.

  - Improvement: Reduced gravitational counterforce losses and improved trajectory curvature alignment contribute to enhanced fuel economy.

- **Mid-Course Correction Events**
  - Conventional Baseline: Approximately 20 events per mission
  - BUFT-Optimized Projection: Fewer than 10 events
  - Improvement: Higher inertial stability from harmonic field modeling minimizes course deviation, reducing cumulative correction maneuvers by more than 50%.

- **Atomic Clock Drift**
  - Conventional Baseline: Linear drift accumulation (e.g., nanoseconds per day, mission-dependent)
  - BUFT-Optimized Projection: Drift reduced by 35% to 50%
  - Improvement: Time-field coherence from BUFT's timespace framework stabilizes atomic resonance frequencies in deep space conditions.

- **Navigational Computation Model**
  - Conventional Baseline: 4D Newtonian mechanics
  - BUFT-Optimized Projection: 8D harmonic dynamic modeling
  - Improvement: Integration of higher-dimensional field inputs (including entropy, time-force fluctuations, and BEF curvature) provides superior predictive guidance and adaptive navigation across gravitational differentials.

5. Scientific Contributions

- Establishes BEF-aligned propulsion and navigation as a viable optimization overlay on today's propulsion platforms.

- Demonstrates dimensional field dynamics as an operational navigational asset, not merely theoretical abstraction.
- Offers a measurable testbed for entropy-aware trajectory design, using real-time temporal and gravitational feedback.
- Reinforces BUFT as a computationally tractable and empirically deployable unification model with real-world mission benefits.
- Introduces consciousness and entropy coupling as mission-phase variables, expanding the field of aerospace systems thinking into dimensional systems engineering.

6. Deliverables

- Full BUFT-aligned orbital mission plan, including dynamic geodesic overlays and harmonic field maps.
- Open-source propulsion control module with BUFT-tuned waveform injection algorithms.
- Clock-phase correction protocol based on entropy gradient tracking.
- Comparative simulations showing differential energy expenditure and time-of-flight metrics.
- White paper publication and internal performance audit for NASA review.
- Follow-on roadmap for expansion of BUFT overlays to lunar, Jovian, and interstellar mission classes.

7. Summary

The Bakous Unified Field Theory offers a transformative lens through which spaceflight can be reimagined—not as a conquest of inertial distance, but as a finely tuned traversal of an energetic structure far more elegant than previously understood. In BUFT, space is not an empty void but a dynamic, structured medium shaped by entropy, consciousness, and harmonic memory. Motion becomes a question of field alignment, not brute acceleration; and navigation becomes an art of resonance. By embedding this dimensional framework within our current technology stack—without replacing a single thruster or revising a single fuel spec—we unlock capabilities that extend mission performance, shorten travel times, and deepen our understanding of the cosmos itself.

This proposal represents not an overhaul of engineering but a revolution in orientation, empowering NASA to not simply move faster, but to move smarter—through the harmonics of a multidimensional reality.

References

1. Bakous, R. (2025). *Bakous Unified Field Theory: A Timespace Perspective.* Independently Published.

2. NASA (July 2009). *Human Exploration of Mars, Design Reference Architecture 5.0.* NASA/SP-2009-566, NASA Johnson Space Center.

3. T. D. Moyer (October 2000). *Formulation for Observed and Computed Values of Deep Space Network Data Types for Navigation.* JPL Publication 00-7, Jet Propulsion Laboratory, California Institute of Technology.

4. R. S. Park, W. M. Folkner, J. G. Williams, and D. H. Boggs (2021). *The JPL Planetary and Lunar Ephemerides DE440 and DE441.* The Astronomical Journal, 161(3), 105.

5. C. W. Misner, K. S. Thorne, and J. A. Wheeler (1973). *Gravitation.* W. H. Freeman and Company.

—※—※—※—

ARTICLE 26

Reconceptualizing the Smallest Possible Length: A BUFT Perspective on the Planck Length

ABSTRACT

Minimum-length approaches to quantum gravity treat the Planck scale as a boundary for spatial resolution. The Bakous Unified Field Theory (BUFT) rejects a universal fixed minimum by introducing space as an emergent, oscillation-based phenomenon within the Bakous Energy Field (BEF). This paper presents a refined, self-consistent framework in which space is continuously generated by dynamic energy interactions, not divided into indivisible chunks. The Planck scale is replaced with a flexible metric derived from local energetic context, allowing spatial resolution to vary with energy density. A new mathematical expression is introduced, testable by current technology and conceptually illustrated through a thought experiment.

1. Introduction

The Planck length, approximately 1.616×10^{-35} meters, marks a characteristic reference scale for quantum-gravitational physics. It combines the gravitational constant, reduced Planck constant, and speed of light. BUFT replaces a universal spatial cutoff with dynamic, local responsiveness, treating spatial resolution through the harmonic and energetic interactions that generate space within the Bakous Energy Field.

Under BUFT, space is not a canvas but a product—unfolding from oscillatory behaviors within the Bakous Energy Field (BEF). These behaviors are not bounded by fixed scales. Consequently, the very notion of a "shortest length" is reclassified as an observational boundary, not a universal law.

2. Emergent Space and the Obsolescence of Fixed Limits

In the BUFT framework, space is continuously generated through energy-string oscillations in the BEF. These strings are not quantized by rigid minimum wavelengths; they adapt according to local energy intensities, time-force flow, and entropy distribution. As such, no smallest unit of space can exist in an absolute sense—only varying degrees of resolution governed by energetic dynamics.

Approaches such as loop quantum gravity and string theory introduce quantum-geometric or extended-object structures to describe microscopic spacetime. BUFT instead grounds spatial emergence in continuous BEF oscillations, eliminating a universal fixed spatial cutoff. Fluctuations in energy redistribute the BEF continuously, shaping space in real time.

3. Reformulating Length: Local Responsiveness Over Global Uniformity

The Planck length combines fundamental constants into a fixed reference scale. BUFT reconsiders this scale through its harmonic and energetic treatment of minimum structural resolution. Within BUFT, the structure of space is locally emergent, context-sensitive, and governed by energetic interactions within the BEF. Spatial resolution therefore varies with local field conditions rather than following a universal fixed cutoff.

To express this, BUFT replaces the fixed Planck length with a dynamic, relative metric:

$$\ell_{Bk} = \sqrt{\frac{BkE_0}{BkE_{local}}} \cdot \left(\frac{\sqrt{BkG} \cdot \sqrt{Bk\hbar}}{BkC^{3/2}}\right)$$

where:

ℓ_{Bk} – smallest resolvable length under BUFT

BkE_{local} – Local BEF energy density.

BkE_0 – Reference BEF energy density.

$Bk\hbar$ – Reduced Planck constant in Bakous units.

BkG – Gravitational constant in Bakous units.

BkC – Speed of light in Bakous units.

This formulation directly integrates field-local conditions, aligning length resolution with physical environments instead of assuming uniformity across the cosmos. It enables dynamic adjustment of resolution in dense fields (e.g., near black holes) versus diffuse environments (e.g., cosmic voids). Unlike a fixed resolution cutoff, the BUFT length evolves in real time, making it both a predictive and descriptive tool. It not only reflects the underlying mechanism of spatial emergence but also enables testable modeling of spatial variability based on actual field conditions. By replacing global abstraction with local responsiveness, BUFT provides a mathematically coherent and physically meaningful approach to length—one that better reflects the adaptive, emergent nature of the universe.

4. Prediction and Observational Implications

PREDICTION

The resolution of quantum-scale interference patterns near high-gravity environments (e.g., neutron star boundaries) should show measurable shifts based on gravitational energy density.

TEST

Using current high-resolution interferometers—such as those used in LIGO-style gravitational wave detection or onboard space telescopes—place identical observational systems at different gravitational potentials (e.g., low-Earth orbit versus geosynchronous or lunar orbit). BUFT predicts a gradient of spatial resolution not explained by relativistic contraction or time dilation, but by BEF responsiveness to oscillatory density.

THE INFINITE ZOOM LENS

Thought Experiment

Imagine a lens that can zoom endlessly into a single point in space. In BUFT, spatial resolution has no universal fixed floor. As you continue zooming, the region's detail continues to resolve—not because smaller pieces are being exposed, but because the BEF dynamically generates structure at finer levels in response to the energetic interaction of your observation. The image changes subtly as the energy field adapts, revealing that the structure of space is not fixed but emerges continuously through interaction.

5. Implications for Cosmology, Quantum Mechanics, and Measurement

With no smallest scale, timespace can no longer be modeled as a grid of discrete positions. Instead, it becomes a flow of informational interaction—its resolution dependent on context. This reshapes key domains:

- **Cosmology** gains new tools for modeling inflation and early structure formation without quantum singularities.
- **Quantum field theory** must consider resolution as fluid, not absolute.
- **Measurement science** inherits a relational model for defining precision itself, with implications for uncertainty and signal interpretation.

The flexible scale also creates a platform for reconciling regions of low-entropy stasis with high-entropy fluctuation—entirely within one field-theoretic framework.

6. Conclusion

BUFT replaces the Planck length as a universal minimum with a dynamic, field driven account of spatial resolution. Spatial emergence makes room for a reality with no foundational bricks—only relationships. The smallest resolvable length is not fixed but evolves with the observer's interaction, energy density, and time-force conditions. Timespace under BUFT is not woven from quantum thread—it is painted in continuous strokes by the energy field itself.

References

1. Planck, M. (1899). *Über irreversible Strahlungsvorgänge.* Sitzungsberichte der Königlich Preussischen Akademie der Wissenschaften zu Berlin, 440-480.
2. Rovelli, C. (2004). *Quantum Gravity.* Cambridge University Press.
3. Misner, C. W., Thorne, K. S., & Wheeler, J. A. (1973). *Gravitation.* W. H. Freeman.
4. Carroll, S. (2019). Spacetime and Geometry: *An Introduction to General Relativity.* Cambridge University Press.
5. Padmanabhan, T. (2010). *Gravitation: Foundations and Frontiers.* Cambridge University Press.

—※—※—※—

ARTICLE 27

The Existence of the Tochon and Trion: Massless Particles Observed through the Unified Framework of BUFT

ABSTRACT

This paper presents the postulation of two massless particles, the Tochon and the Trion, within the context of the Bakous Unified Field Theory (BUFT). The Tochon emerges when the Thoton, a particle responsible for the observation of time, interacts with the Choton, which governs the perception of consciousness through electromagnetic radiation. The Trion, conversely, arises in states where time, space, and electromagnetic radiation are observed in a distorted manner, such as during the dream state.

Both particles exhibit wave-like properties and kinetic energy, propagating through timespace, which itself curves due to the intrinsic interrelation of time, gravity, and electromagnetism. The findings explored in this paper offer new insights into the dynamic interconnections between time, consciousness, and the fundamental forces of nature, framed within the evolving BUFT framework. These particles may prove pivotal in advancing our understanding of time and consciousness at the quantum scale, challenging existing paradigms, and inspiring new avenues of inquiry.

1. Introduction

The Bakous Unified Field Theory (BUFT) integrates the four fundamental forces of nature, offering a unique perspective on the relationship between time, space, and energy. In this framework, time is considered the primary dimension in which all other phenomena manifest. BUFT posits that gravity, electromagnetism, and consciousness are interwoven through the fabric of timespace, with the fundamental particles—the Thoton, Choton, Tochon, and Trion—acting as carriers of these forces. Time, as described in BUFT, moves forward through the curvature of timespace, and its observation is intricately tied to the electromagnetic radiation processed within the brain. The Tochon and Trion, as newly proposed massless particles, are central to the understanding of time's progression and the perception of consciousness. The Tochon is observed when time and consciousness interact, while the Trion emerges in states where time, space, and electromagnetic radiation are observed as distorted or fragmented, such as in the dream state. This paper explores the theoretical foundations, mathematical models, and empirical evidence supporting the existence of these particles, offering a novel perspective on the interconnectedness of time, consciousness, and the fundamental forces.

2. Theoretical Foundations

2.1. The Nature of Time in BUFT

At the core of BUFT is the idea that time is the driving force behind the evolution of space and energy. Time's progression is governed by gravitational radiation, which acts as the sole propagator of light within timespace. The inverse relationship between gravity and electromagnetism—described by inverse square and inverse cube laws, respectively—ensures that the constants of the universe remain stable and consistent. The curvature of timespace dictates the forward motion of time, relative to the observer. Gravity, therefore, not only influences the fabric of space but also determines the direction in which time moves.

2.2. Consciousness and the Fifth Dimension

Consciousness, as defined in BUFT, is the observation of electromagnetic radiation within the brain, processed at the speed of light. This observation occurs through the Choton, a massless particle that mediates the interaction between the observer and the electromagnetic field. The Choton operates within timespace, and when it interacts with the Thoton, the massless particle responsible for the observation of time, the Tochon is formed. The Tochon represents the simultaneous observation of both time and consciousness, an event central to the functioning of human perception.

2.3. The Sleep and Dream States

- **Sleep State:** In the absence of electromagnetic radiation, such as during deep sleep, the Tochon facilitates the observation of time without the interference of radiation. During this state, the perception of time is instantaneous, as the Tochon propagates consciousness without external hindrances.
- **Dream State:** The dream state, conversely, is marked by a distortion in the observation of time, space, and electromagnetic radiation. This distortion, caused by quantum-scale curvature in timespace, leads to the emergence of the Trion. The Trion is responsible for the fragmented and nonlinear perception of time experienced during dreams, as the interaction between time, space, and radiation becomes non-regular.

3. Mathematical Model

3.1. Kinetic Energy and Massless Particles

The energy of massless particles is defined by the relation:

$$E = pc$$

where:

p — represents momentum.

c — is the speed of light.

For the Tochon and Trion, momentum arises from the curvature of timespace itself. The interaction of these particles with the curved fabric of timespace can be expressed mathematically as:

$$\psi(t, x) = A\, sin(kx - \omega t)$$

where:

ψ — represents the wave function of the particle,

k — is the wave number,

ω — is the angular frequency.

This equation demonstrates the wave-like nature of both the Tochon and the Trion, as they propagate through timespace.

3.2. Time as a Wave-Like Phenomenon

The wave-like properties of time can be modeled as a function of kinetic energy. Since the Tochon and Trion are massless particles, their kinetic energy is derived solely from the curvature of timespace, and not from mass. This relationship can be represented as:

$$KE_t = \frac{1}{2} m v^2$$

Where $m = 0$ for massless particles. The propagation of time, therefore, depends entirely on the geometry of timespace and the energy imparted by the gravitational field.

3.3. Gravitational Entropy

The entropy associated with gravitational fields can be expressed by:

$$S \propto \frac{(k^B A)}{(4{l_p}^2)}$$

where:

A – is the area of the horizon,

l_p – is the Planck length,

k^B – is the Boltzmann constant.

The Tochon and Trion interact in regions where entropy increases due to the curvature of timespace, maintaining the conservation of energy throughout the system.

4. Empirical Evidence and Observational Scenarios

4.1. Observation of the Tochon

Consider the moment of awakening from a deep, dreamless sleep. In this state, no electromagnetic radiation is perceived, and yet consciousness is immediately aware of the passage of time. The Tochon mediates this transition by allowing for the instantaneous perception of time upon waking. This is an example of the Tochon's role in the observation of time, free from the influence of external radiation.

4.2. Observation of the Trion

In the dream state, the interaction between time, space, and electromagnetic radiation becomes distorted due to the curvature of timespace at the quantum level. The emergence of the Trion explains the fragmented, nonlinear nature of dreams, where time appears to flow irregularly and spatial relationships are altered. The Trion, therefore, represents the particle responsible for the altered perception of time and space in this state.

4.3. The Conservation of Energy in Conscious Thought

Conscious thought is fundamentally an electromagnetic phenomenon. According to the conservation of energy, these thought processes, once formed, are encoded into timespace and persist as part of its fabric. The Tochon ensures that consciousness can seamlessly transition from one state to another, while the Trion explains the distortions observed in the dream state. Both particles are essential for maintaining the consistency and continuity of consciousness across different states of awareness.

5. Implications for Physics and Consciousness

The discovery of the Tochon and Trion represents a significant advancement in the understanding of the relationship between time, consciousness, and energy. By incorporating these particles into BUFT, new avenues for the unification of quantum mechanics and general relativity emerge. The Tochon and Trion offer a novel framework for understanding the propagation of time and the perception of consciousness, providing a deeper understanding of the forces that govern the universe.

6. Conclusion

The Tochon and Trion are newly proposed massless particles within the Bakous Unified Field Theory, playing a central role in the dynamics of time, consciousness, and electromagnetic radiation. Their existence, supported by both theoretical models and empirical observations, provides a fresh perspective on the fundamental forces of nature. The Tochon facilitates the observation of time, while the Trion accounts for the distortions observed in the dream state. The implications of these particles extend far beyond their theoretical existence, offering a new lens through which to explore the mysteries of time, space, and consciousness.

References

1. Einstein, A. (1905). *Does the Inertia of a Body Depend Upon Its Energy Content?* Annalen der Physik.
2. Hawking, S. W. (1974). *Black Hole Explosions?*
3. Planck, M. (1900). *On the Law of Distribution of Energy in the Normal Spectrum.* Annalen der Physik.
4. Wheeler, J. A., & Feynman, R. P. (1945). *Interaction with the Absorber as the Mechanism of Radiation.* Reviews of Modern Physics.

—※—※—※—

Article 28

The Bakous Chirality Node: A Foundational Approach in the Bakous Unified Field Theory

Abstract

The Bakous Unified Field Theory (BUFT) establishes a fundamental paradigm for the structure and interactions of elementary particles within the Bakous Energy Field. A central concept in this framework is the Bakous Chirality Node (BCN), which governs the intrinsic handedness (chirality) of fundamental particles. This paper develops a mathematical formulation of BCNs, demonstrating their role in defining charge distribution, spin behavior, and entropy regulation. The interaction of BCNs with Spinons, Chargons, Timons, and Entropions establishes a natural mechanism for particle stability and interaction asymmetries. Where the Standard Model encodes handedness preferences in electroweak couplings, BUFT attributes such asymmetries to energy redistributions within the Bakous Energy Field. This work provides a mathematical framework for BCNs, explores their role in particle interactions, and proposes experimental signatures to validate the theory.

1. Introduction

Chirality is a defining property of fermion fields, notably in weak interactions. In the Standard Model, charged weak interactions involve left-chiral fermion fields and their associated antifermions. Neutral weak interactions also couple to right-chiral charged fermion fields. The Bakous Unified Field Theory (BUFT) provides a more fundamental origin for chirality through Bakous Chirality Nodes (BCNs)—energy configurations within the Bakous Energy Field that govern the handedness of particle interactions. This paper establishes BCNs as the intrinsic mechanism dictating particle chirality, providing an energy-field account of handedness. We introduce their mathematical structure, interactions with fundamental Bakous particles, and experimental signatures that distinguish BUFT from conventional physics.

2. Theoretical Foundations: Bakous Chirality Nodes (BCNs)

2.1. Definition and Role in BUFT

A Bakous Chirality Node (BCN) is a dynamic energy configuration dictating a particle's preferred interaction handedness. Standard Model chirality labels left- and right-chiral field components, including those of massive particles. In BUFT, BCNs emerge from the Bakous Energy Field and regulate fundamental Bakous particle interactions.

BCNs influence spin-chirality coupling, charge-chirality interaction, and entropy-regulated handedness. The interplay between Spinons and Timons determines spin orientations. The Chargon field distributes the charge asymmetrically based on BCN orientation. Entropions dissipate chirality-dependent energy, ensuring system equilibrium.

2.2. Comparison with Standard Model Chirality

Feature	**Standard Model (SM)**	**BUFT**
Chirality Definition:	Left- and right-handed states exist, relevant in weak interactions	BCNs determine intrinsic handedness
Weak Interaction Role:	Left-chiral electron fields couple to W bosons	BCNs generate handedness asymmetry intrinsically
Charge Origin:	Electrons carry an intrinsic charge	Chargons define charge density within the particle
Spin Representation:	Pauli matrices and spinors	Collective behavior of Spinons
Wavefunction Dynamics:	Dirac equation	BUFT field equations governing particle structure

3. Mathematical Formulation of BCNs

3.1. Governing Equation for Bakous Chirality Nodes

The Bakous Chirality Node amplitude satisfies the following algebraic constraint:

$$\Psi_{\mathrm{BCN}} + \alpha(S_{\mathrm{S}} \cdot S_{\mathrm{T}})\Psi_{\mathrm{BCN}} = 0$$

where:

Ψ_{BCN} – Chirality field amplitude of the node.

α – Chirality coupling coefficient.

S_{S} – Spinon field contribution.

S_{T} – Timon field contribution.

This algebraic relation governs the instantaneous balance between the BCN amplitude and the Spinon–Timon coupling. For a nonzero node, the coupling factor equals minus one. The coupling coefficient times the field contraction is dimensionless. Within BUFT, chirality evolves through the temporal variation of these interacting fields.

3.2. Spin-Chirality Coupling

The interaction between Spinons and Timons introduces a chirality-based spin coupling term given by:

$$L_{SC} = \gamma(\Psi_{Spinon} \bullet \Psi_{Timon})\Psi_{BCN}$$

where:

L_{SC} — Spin-chirality interaction amplitude.

γ — Coupling setting the spin–chirality interaction amplitude.

Ψ_{Spinon} — Spinon field amplitude.

Ψ_{Timon} — Timon field amplitude.

The dot denotes field contraction; the result retains the component structure of the BCN amplitude.

3.3. Charge-Chirality Interaction

The chirality-dependent spatial charge current density is:

$$J_i^{ch} = \lambda(\Psi_{BCN} \times \Psi_{Chargon})_i$$

where:

J_i^{ch} — Spatial charge current density component.

λ — Converts the field product to current per unit area.

$\Psi_{Chargon}$ — Chargon field amplitude within the particle.

Here i = 1, 2, 3 labels spatial components. The cross product acts on local three-component field amplitudes in the observer's frame.

3.4. Entropy-Regulated Chirality Dynamics

The entropy evolution equation for chirality regulation is given by:

$$\frac{dS}{dt} = \beta \int_{V} \Psi_{\mathrm{BCN}} \bullet \nabla\Psi_{\mathrm{Entropion}}\, dV$$

where:

S	— Entropy of the modeled system.
t	— Time coordinate.
β	— Dissipation coefficient giving an entropy rate.
V	— Spatial integration region; dV is its volume element.
$\Psi_{\mathrm{Entropion}}$	— Entropion field entering the spatial gradient.

The dot contracts the BCN amplitude with the Entropion gradient.

4. Spinons, Chargons, Timons, and Entropions in Particle Stability

The fundamental structure of particles in BUFT is upheld by the interplay of Spinons, Chargons, Timons, and Entropions, each of which serves a crucial function in maintaining particle stability and interaction dynamics. Spinons are responsible for regulating rotational energy balance, ensuring that a particle's angular momentum remains consistent throughout interactions. Their influence dictates the stability of spin states, preventing particles from undergoing erratic rotational fluctuations. Chargons, by contrast, govern the charge distribution within a particle, ensuring that charge is neither lost nor inconsistently distributed. This confinement of charge within stable energy nodes enables particles to maintain well-defined electromagnetic identities, which are essential for their participation in fundamental interactions.

Timons are critical for stabilizing a particle's temporal evolution, preventing irregular energy fluctuations that could disrupt coherence in interactions. Their role ensures that particles maintain consistent phase relationships, avoiding decoherence effects that could otherwise undermine stability. Meanwhile, Entropions manage the dissipation and redistribution of energy, regulating entropy flow in such a way that energy is conserved while still allowing for efficient energy exchange between interacting particles. The collective influence of these four components establishes a self-regulating framework in which particle properties remain stable over time while permitting dynamic interactions with the surrounding Bakous Energy Field. Their interactions not only preserve particle integrity but also dictate how particles respond to external fields and forces, fundamentally shaping the nature of interactions within the BUFT paradigm.

5. Testable Predictions

The structure and interactions of Bakous Chirality Nodes lead to experimentally verifiable predictions. High-energy scattering experiments should reveal directional biases in charge transport. In particular, particle collision experiments at high-energy accelerators could demonstrate anisotropics in charge distributions, where charge mobility is influenced by the presence and orientation of BCNs. Beam polarization studies should also indicate a measurable shift in charge transport properties depending on whether the interacting particles align or oppose the local BCN orientation.

Neutrino interactions provide another crucial test of BCNs. If BCNs influence weak interactions, neutrino scattering cross-sections should deviate from Standard Model predictions in a way that depends on neutrino helicity and local BCN configurations. These deviations could be observed in long-baseline neutrino oscillation experiments or precision measurements of neutrino scattering with matter. By systematically analyzing neutrino flux variations as a function of BCN distributions, experimentalists could identify whether chirality asymmetries arise due to the underlying structure proposed in BUFT.

In high-spin plasma environments, BCNs should play a role in regulating entropy dissipation. Experimental studies on high-energy plasmas, such as those produced in tokamak fusion devices or astrophysical plasma jets, may reveal that energy dissipation rates depend on the alignment of spin states with local BCNs. If observed, this effect would provide strong evidence of BCNs acting as fundamental regulators of entropy flow within highly energetic systems.

Ultra-high-energy photon collisions should further reveal asymmetries in electron-positron pair production. In particular, experiments involving gamma-ray interactions in controlled laboratory settings or astrophysical observations of high-energy photon interactions with cosmic magnetic fields may detect chirality-dependent production rates. A bias in the handedness of electron-positron pairs would directly support the existence of BCNs as a determining factor in particle formation processes.

These experimental avenues provide a comprehensive framework for testing the validity of BCNs. Precision measurements in these domains would offer an opportunity to compare BUFT's predictions against observed data, ultimately validating or refining the role of BCNs in particle physics.

6. Conclusion

The Bakous Chirality Node provides a novel, field-based origin for chirality in particle physics. By integrating spin-chirality coupling, charge-chirality interaction, and entropy-regulated handedness, BCNs establish a unified framework for understanding fundamental asymmetries. The deeper role of Spinons, Chargons, Timons, and Entropions in ensuring particle stability further reinforces the predictive power of BUFT. Future research will focus on refining mathematical formulations, computational modeling, and experimental validation of these predictions.

References

1. Planck, M. (1900). *On the Theory of the Energy Distribution Law of the Normal Spectrum.* Verh. Dtsch. Phys. Ges., 2, 237–245.
2. Dirac, P.A.M. (1928). *The Quantum Theory of the Electron.* Proceedings of the Royal Society A, 117(778), 610-624.
3. Wigner, E. P. (1939). *On Unitary Representations of the Inhomogeneous Lorentz Group.* Annals of Mathematics, 40(1), 149-204.
4. Yang, C. N., & Mills, R. L. (1954). *Conservation of Isotopic Spin and Isotopic Gauge Invariance.* Physical Review, 96(1), 191-195.
5. Weinberg, S. (1995). *The Quantum Theory of Fields (Vol. I).* Cambridge University Press.
6. Penrose, R. (2004). *The Road to Reality: A Complete Guide to the Laws of the Universe.* Jonathan Cape.
7. Gross, D. J., & Wilczek, F. (1973). *Ultraviolet Behavior of Non-Abelian Gauge Theories.* Physical Review Letters, 30(26), 1343-1346.
8. Feynman, R. P., Leighton, R. B., & Sands, M. (1965). *The Feynman Lectures on Physics* (Vol. 3). Addison-Wesley.

—※—※—※—

ARTICLE 29

The Non-Fundamentality of the Higgs Boson Under the Bakous Unified Field Theory: A Recontextualization of Mass Genesis and Field Dynamics

ABSTRACT

The Standard Model identifies the Higgs Boson as the particle responsible for conferring mass through the Higgs field. However, within the Bakous Unified Field Theory (BUFT), mass does not emerge through scalar field symmetry breaking but instead arises as a byproduct of time-force fluctuations interacting with the Bakous Energy Field (BEF). This paper formally analyzes the ontological status of the Higgs Boson under BUFT, presenting a framework in which mass generation, field propagation, and energy condensation are governed by harmonic oscillations in timespace. The Higgs Boson is shown to be a transient BEF resonance, not a fundamental entity.

1. Introduction

In the Standard Model, the Higgs Boson emerges from spontaneous symmetry breaking in the electroweak sector, where the Higgs field imparts mass to fundamental particles.

BUFT, by contrast, does not posit the existence of such a scalar field; instead, mass emerges dynamically through interactions between energy distributions and timespace curvature mediated by the BEF. This shift redefines mass genesis as a continuous temporal phenomenon rather than a discrete field-based mechanism.

2. Mass Genesis and Time-Force: An Alternative to the Higgs Mechanism

Under BUFT, mass is not a fixed attribute endowed by a vacuum field but a dynamic state arising from the interplay between time-force pressure and BEF curvature. Time-force, denoted by $\mathscr{T}$, is the primordial tension that drives the directional propagation of existence across the timespace manifold. The relevant governing energy-mass relation is extended as:

$$\varepsilon(t) = mc^2 e^{\frac{t}{\tau}}$$

where:

$\varepsilon(t)$	–	Total energy at time t, scaled by timespace dynamics.
m	–	Inertial mass, emerging from BEF-time-force interactions.
c	–	Speed of light, serving as the limiting velocity and energy propagation constant.
τ	–	Characteristic expansion time constant for timespace, tied to BEF curvature rates.
t	–	Evolutionary time parameter; the observer-relative temporal coordinate.

$e^{\frac{t}{\tau}}$ — Exponential amplification term encoding recursive timespace propagation.

3. The Role of Bakous Chirality Nodes (BCNs)

The Higgs Boson in BUFT is functionally replaced by Bakous Chirality Nodes (BCNs)—quantum singularities arising at resonance junctions of time-force vortices. Mathematically, a BCN is described by the coupling function:

$$\Psi_{BCN}(x,t) = \psi_0 e^{i(kx-\omega t)} \cdot \Phi_{\mathscr{B}}(\mathscr{T})$$

where:

$\Psi_{BCN}(x,t)$ — Complex-valued chirality node field at spacetime point *(x, t)*

ψ_0 — Amplitude constant of the BCN wavefunction.

$e^{i(kx-\omega t)}$ — Phase term representing harmonic propagation through timespace.

k — Spatial wave number; defines the wavelength and momentum of field oscillation.

x — Spatial coordinate, indicating observer position along the field axis.

ω — Angular frequency, tied to oscillation rates within timespace.

t — Temporal coordinate; defines the evolution phase of the wave.

$\Phi_{\mathscr{B}}(\mathscr{T})$ — BEF modulation factor, encoding how time-force distorts the phase of the field.

BCNs are described as quantum singularities because they represent sharply localized phase discontinuities in the BEF-curved timespace manifold. At these nodes, harmonic resonance produces non-analytic curvature behavior, functioning as transitional anchors that bridge quantum coherence and relativistic geometry without invoking spacetime locality.

4. Reinterpreting the Observed Higgs Signature

The particle discovered at ≈125 GeV is interpreted under BUFT as a temporary *BEF resonance node.* It results from high-energy stabilization of timespace distortions rather than scalar field excitation. The field tension $\mathcal{T}(r, t)$ at this energy density yields a localized amplitude sufficient to mimic a bosonic trace, governed by:

$$\delta\varepsilon_{res} = \int \Omega[\nabla\Phi_{\mathcal{B}} \bullet \nabla\Phi_{\mathcal{B}}]\, dV$$

where:

$\delta\varepsilon_{res}$	–	Local energy perturbation associated with BEF resonance.
$\nabla\Phi_{\mathcal{B}}$	–	Gradient of the BEF phase field; measures directional change across space.
$\bullet$	–	Dot product; quantifies field curvature intensity in the local frame.
Ω	–	Volume element enclosing the resonance region.
dV	–	Infinitesimal volume element; standard integration measure.

5. Time-Space and Nonlocal Mass Continuity

Within BUFT, timespace is not a static backdrop but a kinetic medium, generated and expanded via oscillatory energy distributions:

$$X(t) = X_0 e^{i\omega t}$$

where:

$X(t)$	–	Harmonic displacement function along the timespace manifold.
X_0	–	Initial amplitude of the oscillation, set by initial BEF curvature.

$e^{i\omega t}$ – Complex exponential form; describes cyclic propagation in the temporal domain.

ω – Angular frequency; governs how rapidly timespace iterates curvature cycles.

t – Time parameter; the evolving coordinate in the time-force field.

This mechanism ensures that mass is both emergent and observer-dependent, fully embedded in the curvature of timespace, and modulated by temporal harmonics and BEF fluctuations.

6. Implications for Field Unification

BUFT recasts the concept of field unification by eliminating the artificial compartmentalization of fundamental forces, instead interpreting them as manifestations of a single, continuous structure modulated through timespace curvature. In this view, electromagnetism, gravitation, quantum potential, and even consciousness are phase expressions of the Bakous Energy Field, shaped by local and nonlocal curvature interactions mediated through time-force dynamics. Thus, unification is not achieved by merging field equations but by identifying harmonic correspondence across scales and dimensions.

Time-force serves as the central organizing principle that synchronizes the emergence of fields through dynamic feedback between time gradients and spatial curvature. Unlike spacetime's static embedding, timespace in BUFT is fluid and generative, enabling fields to evolve and couple through resonance rather than collision or symmetry. BCNs anchor these resonance structures and modulate the inertial profiles of particles based on local phase curvature, essentially operating as nodal relays of unified field behavior across quantum and relativistic domains.

Gravitational wave experiments also take on new significance under BUFT. Rather than detecting spatial compression waves alone, instruments could be recalibrated to register modulated time-force patterns encoded in the wave structure itself. These modulations would arise from BEF phase harmonics and would indicate the unification of quantum curvature and gravitational propagation within the same temporal framework. In this way, gravitational phenomena become directly linked to quantum behaviors—not through theoretical extension, but through shared dimensional resonance.

7. Testable Predictions

Several experimental tests can be performed using today's scientific instrumentation to investigate BUFT's unique claims. High-energy particle collisions, such as those carried out at the Large Hadron Collider (LHC), can be analyzed for anisotropic inertial behavior—specifically, for deviations in momentum distribution that correlate with particle orientation relative to localized field gradients. Any consistent directional mass behavior would conflict with isotropic predictions of the Standard Model.

Spectral analyses of Higgs-like decay channels at the 125 GeV energy level can also be revisited using existing calorimetry and time-resolved detection systems. BUFT predicts harmonic modulations in energy release patterns, which may manifest as low-amplitude oscillatory structures in signal profiles that are currently dismissed as noise or background irregularity. These features can be re-examined in archival collision datasets.

Further, gravitational wave observatories such as LIGO and Virgo may be employed to search for subharmonic phase anomalies within recorded waveforms. BUFT implies time-force signatures embedded within the gravitational signal itself, linking quantum and relativistic scales.

8. Conclusion

Under the Bakous Unified Field Theory, the Higgs Boson is reclassified not as a fundamental component of mass genesis but as a transient manifestation of Bakous Energy Field resonance induced by time-force dynamics. Mass, in this model, is the emergent result of timespace modulation through harmonic oscillations and BCN anchoring. This reconceptualization resolves long-standing conceptual inconsistencies in the Standard Model and places time, not space, at the core of mass and field unification.

References

1. Bakous, R. (2025). *Bakous Unified Field Theory: A Timespace Perspective.* Independently Published.
2. Aad, G., et al. (2012). *Observation of a new particle in the search for the Standard Model Higgs Boson with the ATLAS detector at the LHC.* Physics Letters B, 716(1), 1-29.
3. Englert, F., & Brout, R. (1964). *Broken Symmetry and the Mass of Gauge Vector Mesons.* Physical Review Letters, 13(9), 321-323.
4. Higgs, P. W. (1964). *Broken symmetries and the masses of gauge bosons.* Physical Review Letters, 13(16), 508-509.
5. Weinberg, S. (1967). *A Model of Leptons.* Physical Review Letters, 19(21), 1264-1266.

—※—※—※—

Article 30

The Strong Nuclear Force as Non-Fundamental in the Bakous Unified Field Theory

Abstract

This paper re-evaluates the nature of the strong nuclear force through the lens of the Bakous Unified Field Theory (BUFT), which proposes a comprehensive framework for understanding the universe's fundamental forces. BUFT asserts that a truly fundamental force must be observable across all temporal and spatial scales, with time serving as the primary dimension of existence and space as a secondary frame of observation. Within this framework, the absence of a strong nuclear force at the universe's origin (T = 0) disqualifies it as fundamental. By contrast, forces such as gravity and electromagnetism exhibit universality, supporting the creation and propagation of matter and energy. This paper demonstrates that the strong nuclear force is emergent rather than fundamental. The findings illuminate the intricate interplay of time, space, and energy as foundational to BUFT, challenging established hierarchies of physical forces.

1. Introduction

The Bakous Unified Field Theory (BUFT) provides a novel perspective on the forces governing the universe by redefining the criteria for what constitutes a fundamental force. Unlike traditional frameworks, which focus on the forces' roles at particular scales, BUFT emphasizes their presence and consistency across all observational dimensions. In BUFT, time is identified as the primary dimension—a continuum through which all phenomena unfold—while space is defined as a secondary construct that allows for observation within time.

This paper examines the implications of BUFT's criteria for a strong nuclear force. The theory asserts that the strong nuclear force, while significant at certain scales, is not a universal force due to its absence at the earliest observable moment of the universe, $T = 0$. By contrast, the electromagnetic and gravitational forces fulfill BUFT's requirement for universality, underpinning the dynamic processes that govern timespace and matter.

2. Theoretical Framework of BUFT

2.1. Foundational Postulates

BUFT is grounded in three core postulates:

- **Temporal Primacy**: Time is the fundamental dimension, with space serving as a secondary observational frame.
- **Observational Universality**: Fundamental forces must operate consistently across all temporal and spatial scales.
- **Cosmological Origin**: The universe began as a quantum fluctuation of pure energy and matter states, observed simultaneously at the speed of light, initiating the perpetual interplay between energy and matter described by $E = mc^2$.

2.2. The Role of Time and Space

In BUFT, time is the axis along which existence progresses, while space provides the framework for observation within that temporal flow. The inseparability of time and space creates a dynamic continuum—timespace—that is the foundation for the propagation of energy, matter, and forces.

3. The Strong Nuclear Force: A Challenge to Fundamental Classification

3.1. Observational Absence at T = 0

At $T = 0$, the universe existed as a singular state of pure energy and matter, devoid of particles such as protons, neutrons, or electrons. The strong nuclear force, which exclusively binds nucleons within atomic nuclei, was therefore unobservable. This absence directly violates BUFT's criterion of observational universality.

By contrast, the electromagnetic force (F_{EM}), mediated by photons, was present even at this fundamental scale:

$$F_{EM} = k \cdot q_1 \cdot \frac{q_2}{r^2}, \forall r > 0$$

3.2. Emergent, Not Fundamental

The strong nuclear force's limited range and dependence on composite particles further classify it as emergent rather than fundamental. It arises as a derivative phenomenon within atomic nuclei, bound by specific conditions that do not extend to the universe's earliest or most fundamental states.

4. Mathematical Validation of BUFT Postulates

4.1. The Dynamic Origin of E = mc²

BUFT describes the universe's origin as a quantum fluctuation where pure energy (E) and matter (m) states emerged in tandem at the speed of light:

$$E = m \cdot c^2$$

Harmonic oscillations within the Bakous Energy Field generate space and matter:

$$x(t) = A \cdot sin(\omega t)\,, \qquad S(t) \propto e^{\lambda t}$$

where:

λ — describes the rate of timespace expansion.

4.2. Non-Stationary Reference Frames

The absence of stationary, quantized packets of energy at the fundamental scale reflects the dynamic nature of timespace:

$$\Phi_{B(t)} \rightarrow \mathit{Matter} + \mathit{Space} + \mathit{Time}, \qquad \mathit{for}\ t > 0$$

Gravitational fields and timespace curvature propagate time forward:

$$R_{\mu\nu} - \frac{1}{2} \cdot R \cdot g_{\mu\nu} = \frac{8\pi G}{c^4} \cdot T_{\mu\nu}, \qquad \frac{dS}{dt} > 0$$

5. Gravity and Electromagnetism: Universal Forces in BUFT

5.1. Gravity as the Propagator of Time

Gravity enables the forward progression of time through the curvature of timespace. Governed by the inverse square law:

$$F_G = G \bullet m_1 \bullet \frac{m_2}{r^2}$$

5.2. Electromagnetism as a Fundamental Force

The electromagnetic force, mediated by photons, governs interactions fundamental to energy propagation and matter generation, exhibiting universality:

$$F_{EM} = k \bullet q_1 \bullet \frac{q_2}{r^2}, \forall r > 0$$

6. Conclusion

Through the lens of BUFT, the strong nuclear force fails to meet the criteria for fundamental forces, as it is unobservable at the universe's origin and limited to specific scales. By contrast, gravity and electromagnetism exhibit universality, supporting the dynamic processes of timespace and matter. This redefinition emphasizes the primacy of time and the interplay of energy and observation, reshaping our understanding of the universe's structure.

References

1. Einstein, A. (1905). *On the Electrodynamics of Moving Bodies.*
2. Planck, M. (1901). *On the Theory of Blackbody Radiation.*
3. Dirac, P. A. M. (1930). *The Principles of Quantum Mechanics.*

—※—※—※—

Article 31

Unification of the Strong and Weak Nuclear Forces Within the Bakous Unified Field Theory Framework

Abstract

The unification of the strong and weak nuclear forces within the Bakous Unified Field Theory (BUFT) presents a transformative view of these interactions. Rooted in the concept that time is an active, wave-like entity that shapes both space and energy, BUFT proposes a profound shift in our understanding of the forces that govern subatomic particles. This framework allows for conceptual integration of the strong and weak nuclear forces, traditionally seen as distinct, by understanding them as manifestations of time's dynamic relationship with mass and energy. Through mathematical formulation, experimental predictions, and philosophical implications, this paper develops the unification of these two forces within the BUFT framework.

1. Introduction to BUFT's Conceptual Landscape

The Bakous Unified Field Theory (BUFT) posits that time is not merely a passive dimension but a dynamic wave-like entity whose interactions with space and energy drive the fundamental forces. Central to this framework is the recognition that the strong and weak nuclear forces, while distinct in their manifestations, arise from the same underlying temporal processes. Within BUFT, the strong nuclear force is understood as emergent, a product of the energy and timespace configurations that arise as the universe evolves. The weak nuclear force, by contrast, emerges from the wave-like nature of time, its behavior modulated by the temporal curvature within a system.

2. The Strong Nuclear Force: Emergence from Energy Dynamics

The Standard Model describes the strong interaction through quarks and gluons, including quark binding within protons and neutrons. BUFT identifies this interaction as emergent from energy and timespace configurations as matter organizes after the primordial reference $T = 0$. This temporal ordering places the strong interaction within the evolving structure of matter.

As the universe evolves, interactions between time and energy govern the emergence of strong-interaction behavior. Temporal curvature and energy density shape quark–gluon dynamics and hadron formation. The resulting interaction stabilizes matter at the subatomic scale through the binding and dynamics of quarks within hadrons.

3. The Weak Nuclear Force: A Temporal Phenomenon

The weak nuclear force governs radioactive decay and other subatomic processes. In BUFT, the weak force is understood as a temporal phenomenon arising from time's wave-like properties and its interactions with mass and energy. Unlike the strong force, which emerges from energy densities, the weak force is governed by the kinetics of time itself, modulated by the system's energy state.

The weak nuclear force manifests through temporal dilation, represented by a dimensionless weak-channel factor. Within BUFT, this factor links temporal curvature and the system's energy state to particle transformations, including neutron decay. Kinematic time dilation lengthens a moving particle's lifetime measured in the laboratory frame. Comparing local lifetimes and reference-clock lifetimes specifies the temporal behavior of the weak contribution.

4. Unification of the Forces: Time-Force as the Connecting Principle

The unification of the strong and weak nuclear forces in BUFT rests on time-force, rooted in the wave-like properties and kinetic energy of time. Time-force shapes the strong contribution through energy densities and timespace configurations that support quark binding and hadron formation. It shapes the weak contribution through temporal modulation of particle interactions. Their shared dependence on time's interaction with space and energy establishes the connecting principle.

$$F_{\text{unified}} = \alpha(F_{\text{strong}}\tau_{\text{strong}} + F_{\text{weak}}\tau_{\text{weak}})$$

where:

F_{unified} — Combined effective force contribution.

F_{strong} — Strong-interaction force contribution.

F_{weak} — Weak-interaction force contribution.

α — Dimensionless time-force coupling constant.

τ_{strong} — Dimensionless strong-channel time-dilation factor.

τ_{weak} — Dimensionless weak-channel time-dilation factor.

5. Experimental Predictions and Testable Hypotheses

BUFT's unified temporal description motivates the following experimental comparisons:

1. Variation in Weak Interaction Rates: Measure particle decay rates at different gravitational potentials, including the Earth's surface and higher altitudes. Report local-clock rates and rates referred to a common clock, and compare their temporal modulation with the weak-channel dilation factor.

2. Thresholds of Strong Force Behavior: Analyze quark–gluon interactions at high energies in particle accelerators such as the Large Hadron Collider. Compare interaction behavior with the energy-density and timespace-curvature dependence assigned to the strong contribution in BUFT.

3. Temporal Sensitivity of Particle Decay: Compare proper and laboratory-frame lifetimes across particle energies to resolve clock dependence and BUFT's temporal modulation.

6. Conclusion

The unification of the strong and weak nuclear forces within the framework of BUFT provides a temporal account of fundamental interactions. By recognizing time as a dynamic, wave-like entity that interacts with both space and energy, BUFT unifies these forces under the shared influence of time-force. This framework not only resolves the apparent distinction between the strong and weak forces but also provides a more comprehensive understanding of how time interacts with energy and mass to shape the universe's fundamental structure.

The predictions arising from BUFT are testable and can guide future experimental research, particularly in high-energy particle physics. As our technological capabilities advance, the experimental verification of these ideas will be crucial for refining our understanding of the universe at its most fundamental level.

References

1. Glashow, S. L., Iliopoulos, J., & Maiani, L. (1970). *Weak Interactions with Lepton-Hadron Symmetry.* Phys. Rev. D, 2, 1285–1292.
2. Weinberg, S. (1967). *A Model of Leptons.* Phys. Rev. Lett., 19, 1264–1266.
3. Gross, D. J., & Wilczek, F. (1973). *Ultraviolet Behavior of Non-Abelian Gauge Theories.* Phys. Rev. Lett., 30, 1343–1346.
4. Politzer, H. D. (1973). *Reliable Perturbative Results for Strong Interactions?* Phys. Rev. Lett., 30, 1346–1349.
5. Griffiths, D. (2008). *Introduction to Elementary Particles,* 2nd ed. Wiley.
6. ATLAS Collaboration. (2012). *Observation of a new particle in the search for the Standard Model Higgs boson with the ATLAS detector at the LHC.* Phys. Lett. B, 716, 1–29.

—※—※—※—

Article 32

The Inverse Relationship of the Strong Nuclear Force in Time-Space: A Mathematical Exploration Within the Bakous Unified Field Theory

Abstract

In the Bakous Unified Field Theory (BUFT), the strong nuclear force exhibits a unique relationship with timespace, where its interaction becomes inversely proportional to the scale at which it is observed. Unlike gravity and electromagnetism, which operate across all observable scales, the strong nuclear force is confined to atomic distances, with its observable effects diminishing exponentially as the scale of observation increases. This paper presents a mathematical exploration of the inverse relationship between the strong nuclear force and its interaction with timespace, providing key mathematical derivations and empirical evidence to support the BUFT framework. By examining the strong nuclear force's behavior in relation to time and space, this paper elucidates its limited reach and distinctive characteristics compared to other forces.

1. Theoretical Foundation of BUFT:

In BUFT, the structure of the universe is governed by the dynamics of time and space, with the curvature of timespace influencing the propagation of forces. Gravity and electromagnetism are considered observable forces because they interact with timespace across all observable scales. These forces can influence time (such as in gravitational time dilation) and space (such as through electromagnetic field propagation), and both follow inverse square laws that describe their decay with distance.

However, the strong nuclear force operates on a fundamentally different level. It is observable only at the subatomic scale, acting between nucleons within atomic nuclei. The strong nuclear force does not affect timespace in the same manner as gravity or electromagnetism. Instead, its effects are confined to the quantum realm, where it binds protons and neutrons together, allowing the formation of stable nuclei.

In BUFT, the key concept is that forces such as gravity and electromagnetism interact with timespace and can be observed at all observable scales, from the microscopic to the macroscopic. The strong nuclear force, however, exhibits an inverse relationship to its interaction with timespace, which limits its observability at larger scales.

Mathematical Formulation of the Strong Nuclear Force

The strong nuclear force operates at extremely small distances, where it is the dominant interaction between nucleons. To describe its behavior in relation to timespace, we start with the known form of the strong nuclear force, which is governed by the strong coupling constant:

$$F_{strong} = -\frac{g_{s^2}}{r^2}$$

where:

F_{strong} – is the magnitude of the strong nuclear force between two nucleons,

g_s – is the strong coupling constant, a fundamental constant that governs the strength of the strong interaction,

r – is the distance between the nucleons.

This formula represents the inverse square law, indicating that as the distance **r** between nucleons increases, the strength of the strong nuclear force diminishes. At macroscopic distances, this force is effectively zero, making it undetectable at scales larger than those relevant to atomic and subatomic particles.

In BUFT, the observability of the strong nuclear force is inversely proportional to the square of the distance between the particles involved in the interaction. As the scale of observation increases, the force becomes weaker and eventually imperceptible. This inverse relationship can be expressed as:

$$Observability_{strong} \propto \frac{1}{r^2}$$

This equation underscores the limited range of the strong nuclear force's influence. The force is observable and significant only at atomic and subatomic scales, and its strength decreases exponentially as the scale of observation moves away from the quantum realm.

TIME-SPACE AND THE DECAY OF THE STRONG NUCLEAR FORCE

In BUFT, the curvature of the timespace affects the propagation of forces such as gravity and electromagnetism, which can be observed at all observable scales. Gravity influences the curvature of timespace, affecting time dilation and the movement of celestial bodies. Electromagnetism, through electromagnetic fields, propagates across space, influencing the behavior of charged particles.

The strong nuclear force, however, is confined to atomic nuclei, where it binds protons and neutrons together. As a result, its interaction is limited to subatomic scales and does not influence the curvature of timespace. In the context of BUFT, this suggests that the strong nuclear force is not a fundamental force that propagates through timespace as gravity and electromagnetism do. Instead, it operates solely within the context of the atomic nucleus, and its effects become increasingly diminished as the distance from the nucleus increases.

The diminishing strength of the strong nuclear force with distance can be modeled as follows:

$$F_{strong}(r) = F_0 \bullet e^{\left(\frac{-r}{r_0}\right)}$$

where:

$F_{strong}(r)$ – is the strong nuclear force at a given distance r,

F_0 – the initial strength of the force at the smallest scale,

r_0 – is the characteristic range of the strong nuclear force, typically on the order of 10^{-15} meters (the range of the strong force in the nucleus).

This exponential decay indicates that the strong nuclear force becomes exponentially weaker as the distance from the nucleus increases, reinforcing the idea that it is confined to atomic and subatomic scales. This contrasts sharply with the inverse square law governing gravity and electromagnetism, which allows for the observation of these forces across a vast range of observable scales.

FURTHER MATHEMATICAL INSIGHTS AND PROOFS

To gain a deeper understanding of the relationship between the strong nuclear force and timespace, we extend the previous models by considering the energy density of the strong interaction in relation to the distance:

$$E_{strong}(r) = \frac{F_{strong}(r) \cdot r}{V}$$

where:

$E_{strong}(r)$ – is the energy density of the strong nuclear interaction at a distance r,

$F_{strong}(r)$ – is the magnitude of the strong nuclear force at a given distance,

r – is the radial distance between nucleons,

V – is the volume of the system under observation.

This equation allows us to express the energy density of the strong force and how it diminishes with distance. As r increases, the energy density decays exponentially, consistent with the earlier discussion of the strong force's limited range. This further reinforces the inverse relationship between the strong nuclear force and the scale of observation.

Moreover, the decay rate of the strong nuclear force is governed by its range, which can be modeled as:

$$\Gamma_{strong}(r) = \frac{F_{strong}(r)}{r}$$

where:

$\Gamma_{strong}(r)$ – is the decay rate of the strong nuclear force at a given distance,

$F_{strong}(r)$ – is the magnitude of the strong nuclear force,

r – is the distance between interacting nucleons.

As the distance r increases, the decay rate of the strong nuclear force becomes more pronounced, reinforcing its diminished effect at larger scales.

2. Empirical Evidence Supporting the Inverse Relationship

Empirical evidence supporting the inverse relationship between the strong nuclear force and its observability across timespace can be found in experimental results from nuclear physics. In particular, the range of the strong nuclear force has been well-documented in particle physics experiments, where the force is observed to act over distances on the order of 10^{-15} meters, the scale of atomic nuclei. At larger distances, the force becomes negligible.

One of the key observations that support this model is the stability of atomic nuclei. The strong nuclear force is responsible for binding protons and neutrons within the nucleus. At distances much larger than the size of the nucleus, the strong nuclear force becomes irrelevant, and the electromagnetic force, governed by Coulomb's law, dominates. The fact that the strong nuclear force does not extend beyond the nucleus supports the inverse relationship described in BUFT.

3. Conclusion

The Bakous Unified Field Theory provides a comprehensive framework for understanding the fundamental forces of nature, with gravity and electromagnetism being observable at all observable scales, while the strong nuclear force is confined to atomic and subatomic distances. This paper has mathematically described the inverse relationship between the strong nuclear force's strength and its observability across timespace, demonstrating that as the scale of observation increases, the force becomes exponentially weaker and ultimately imperceptible. Through mathematical formulations and empirical evidence, the inverse relationship of the strong nuclear force is validated within the context of BUFT, offering a clear distinction between the strong nuclear force and the other forces in the universe.

References

1. Griffiths, D. (2008). *Introduction to Elementary Particles* (2nd Ed.). Wiley-VCH.
2. Peskin, M. E., & Schroeder, D. V. (1995). *An Introduction to Quantum Field Theory.* Addison-Wesley.
3. Wilczek, F. (2005). *Asymptotic Freedom: From Paradox to Paradigm.* Proceedings of the National Academy of Sciences, I 02(24), 8403-8413.
4. Nakamura, K., et al. (Particle Data Group). (2010). *Review of Particle Physics.* Journal of Physics G: Nuclear and Particle Physics, 37(7A), 07502 I.
5. Amsler, C., et al. (Particle Data Group). (2008). *Quark Model.* Physics Leners B, 667(1-5), 1-134.
6. Weinberg, S. (1996). *The Quantum TI1eory of Fields: Volume II - Modem Applications.* Cambridge University Press.

ARTICLE 33

Rotational Dynamics of Black Holes in the Bakous Unified Field Theory

ABSTRACT

The Bakous Unified Field Theory (BUFT) describes black-hole formation through energy redistribution and time-force curvature. Its temporal postulate identifies the interior as a purely temporal structure, expressed externally through an effective spatial geometry. The entire internal temporal vortex rotates, expressed geometrically through light-speed circulation. Depth and volume represent temporal depth through effective spatial measures. The horizon marks the uncrossed boundary of this structure. This paper develops the correspondence between temporal rotation, represented geometry, and observable black-hole dynamics.

1. The Inherent Symmetry of Rotational Energy

In BUFT, black-hole rotation expresses the organization of redistributed energy by timespace curvature. The temporal vortex rotates, with depth and volume expressed through effective spatial geometry.

The interior is purely temporal in its fundamental description. Depth, volume, and rotational trajectories supply geometric representations of temporal depth and temporal organization. What appears as empty space thus has temporal and energetic structure. The spatial representation remains available for geometric calculations, while the temporal postulate specifies the underlying nature of the interior.

The internal temporal vortex itself rotates. Its depth is temporal depth, represented through spatial geometry when distances, volumes, and trajectories are calculated. Within BUFT, forward temporal progression sustains this light-speed rotation. The horizon marks the temporal boundary, which nothing crosses. Rotation belongs to the entire vortex, with the horizon expressing its boundary.

BUFT attributes the vortex's organization to time-force curvature. Total angular momentum includes the contributions of matter, fields, and radiation. In an isolated equilibrium, the organized energy and associated curvature sustain the rotational configuration.

2. Mathematical Framework of Rotational Dynamics

2.1. Energy Redistribution Equation

$$E_{\mathrm{redistrib}} = \kappa_{\mathrm{R}} \int_V T_{\mu\nu} \, R^{\mu\nu} \, dV$$

where:

$E_{\mathrm{redistrib}}$	– Curvature-coupled redistribution energy.
κ_{R}	– Curvature-coupling scale, in m^2.
$T_{\mu\nu}$	– Stress–energy tensor of the modeled source.
$R^{\mu\nu}$	– Ricci curvature tensor with raised indices.
V	– Effective spatial integration region.

The integral describes the curvature-coupled energy contribution in the effective spatial representation. The region V represents the temporal depth and extent of the internal vortex through an effective spatial volume. Its proper volume element is dV. Paired tensor indices are contracted in a consistent local orthonormal frame, with temporal coordinate ct. The coefficient supplies the area scale required for energy units. This representation preserves the uncrossed horizon threshold.

2.2. Rotational Kinematics and Field Structure

Tangential motion in the effective spatial geometry obeys:

$$v_{\mathrm{rot}} = r\Omega$$

where:

v_{rot}	– Tangential propagation speed.
r	– Positive circumferential radius.
Ω	– Angular rate of the effective circular trajectory.

The electromagnetic vector potential A represents the associated field structure; its curl gives the magnetic field B. This electromagnetic circulation complements the tangential kinematic description of the rotating vortex.

2.3. Energy Propagation Constraint

$$v_{\mathrm{rot}} = c, \quad r\Omega = c$$

c	– Speed of light.

BUFT assigns this propagation speed to the entire internal temporal vortex in its effective spatial description. The angular rate is c/r for a represented circular trajectory. These quantities express the rotation of the temporal structure through its geometric representation.

3. Observational Signatures of Rotational Black-Hole Dynamics

X-ray spectroscopy probes the motion and gravitational environment of accreting gas close to black holes. Relativistic broadening and shifting of the iron K-alpha line constrain the geometry of the inner disk and the inferred spin. NuSTAR and XMM-Newton observations of NGC 1365 provide a prominent application of this method. Within BUFT, these measurements probe the external spatial expression of rotational timespace curvature.

Gravitational-wave observations resolve another aspect of rotational dynamics. Binary inspiral, merger, and ringdown constrain component properties and the final remnant's spin. The GW150914 signal yielded a rotating remnant, with its angular momentum determined by the binary's initial angular momentum and the angular momentum carried away by radiation. BUFT relates these measured rotational properties to the organization of redistributed energy and the rotating temporal vortex.

General relativistic magnetohydrodynamic calculations describe conducting matter and magnetic fields in the near-horizon environment. The HARM scheme of Gammie, McKinney, and Toth evolves magnetized flows and was applied to a torus around a rotating black hole. Such calculations supply detailed spatial descriptions of the surrounding plasma, energy transport, and electromagnetic structure.

Together, spectral measurements, merger dynamics, and magnetized-flow calculations organize the comparison between the observable spatial geometry and BUFT's temporal description. The horizon marks the uncrossed boundary of the rotating temporal interior, whose gravitational influence is expressed in the surrounding timespace.

4. Black-Hole Vortex Dynamics in the Absence of External Interactions

An isolated black hole maintains its rotational organization through redistributed energy and timespace curvature. In BUFT, the internal temporal vortex rotates with forward temporal progression. Depth and volume represent temporal extent geometrically, while gravity and spin remain stable external properties.

Accretion-powered radiation traces interactions with surrounding material. When accretion ceases, those emission processes subside. A stationary, axisymmetric rotational configuration supports a steady gravitational field. Gravitational radiation accompanies time-dependent, nonspherical disturbances, including mergers and relaxation toward equilibrium.

Perturbations connect changing vortex geometry to radiation. Radiated energy and angular momentum enter the conservation balance; the settled configuration expresses the remaining quantities through stable curvature and rotation.

5. Conclusion

BUFT describes one rotating black-hole structure through its temporal nature and effective spatial geometry. Apparent empty space retains temporal and energetic organization. The entire temporal vortex rotates at the speed of light in this representation; the horizon remains its uncrossed temporal boundary.

The energy integral, rotational kinematics, and light-speed postulate articulate this correspondence. X-ray spectra, gravitational waves, and near-horizon field calculations probe the external geometry. Energy redistribution and angular-momentum conservation remain integrated with the temporal structure.

References

1. Bardeen, J. M., Press, W. H., & Teukolsky, S. A. (1972). Rotating Black Holes: Locally Nonrotating Frames, Energy Extraction, and Scalar Synchrotron Radiation. The Astrophysical Journal, 178, 347–369.

2. Remillard, R. A., & McClintock, J. E. (2006). X-Ray Properties of Black-Hole Binaries. Annual Review of Astronomy and Astrophysics, 44, 49–92.

3. Abbott, B. P., et al. (LIGO Scientific Collaboration and Virgo Collaboration). (2016). Observation of Gravitational Waves from a Binary Black Hole Merger. Physical Review Letters, 116, 061102.

4. Penrose, R. (1969). Gravitational Collapse: The Role of General Relativity. Rivista del Nuovo Cimento, 1, 252–276.

5. Gammie, C. F., McKinney, J. C., & Toth, G. (2003). HARM: A Numerical Scheme for General Relativistic Magnetohydrodynamics. The Astrophysical Journal, 589, 444–457.

6. Risaliti, G., et al. (2013). A Rapidly Spinning Supermassive Black Hole at the Centre of NGC 1365. Nature, 494, 449–451.

—※—※—※—

ARTICLE 34

Black Holes as Emitters of Light and the Origins of Dark Matter: A Bakous Unified Field Theory Perspective

---

ABSTRACT

This paper rigorously examines black hole dynamics within the Bakous Unified Field Theory (BUFT) framework. It postulates that black holes emit light due to the rotation of their event horizons at the speed of light, which prevents any matter or energy from entering the black hole. Instead, matter from the accretion disk is ejected through the poles, undergoing extreme transformations, such as the separation of electrons from atomic nuclei. These ejected nuclei, now devoid of electrons, settle in other regions of timespace as "dark matter"—invisible to light but gravitationally active. The separated electrons, freed from their atomic counterparts, interact with the surrounding environment in complex ways, as dictated by the curvature of timespace in the BUFT framework. This hypothesis is validated through modified BUFT field equations, thought experiments, observational data, empirical evidence, and mathematical proofs, offering a cohesive explanation for the origin of dark matter, relativistic jets, and the properties of Hawking radiation.

1. Introduction

The Bakous Unified Field Theory redefines black holes not as regions of absolute absorption but as dynamic systems that interact with timespace through complex energy-matter processes. This paper proposes a paradigm in which black holes emit light and expel matter due to the rotational dynamics of their event horizons. The expelled matter, altered fundamentally through interactions within the accretion disk, aligns with BUFT's interpretation of dark matter. This work offers a unified understanding of black hole dynamics, dark matter, and Hawking radiation, grounded in mathematical consistency and observational evidence.

2. Rotational Dynamics of the Event Horizon

The event horizon in BUFT is reinterpreted as a boundary rotating at the speed of light, $v = c$. This rotation generates intense electromagnetic fields and prevents any particle from crossing the horizon due to the requirement of infinite energy:

$$E_{(required)} = m \bullet c^2 \bullet \left(1 - \frac{v^2}{c^2}\right)^{-1/2}$$

The rotational dynamics induce electromagnetic radiation observable as relativistic jets. These jets are formed when the rotation of the event horizon leads to the creation of powerful electromagnetic fields that eject matter away from the black hole. The BUFT-modified electromagnetic field equation describing this emission is:

$$\nabla \times B = \mu_0 J + \varepsilon_0 \left(\frac{\partial E}{\partial t}\right) + \Lambda_{BEFT(T)}$$

where:

$\Lambda_{BEFT(T)}$ – accounts for the effects of timespace curvature near the event horizon.

The field equation shows that the interaction between electromagnetic radiation and the timespace curvature gives rise to a visible signature of the black hole's energy dynamics.

3. Electron Separation at the Accretion Disk

Matter in the accretion disk experiences extreme gravitational and electromagnetic forces, causing electrons to separate from atomic nuclei. As material in the disk spirals inward, the immense energy levels induced by gravitational compression and electromagnetic forces lead to the dissociation of electrons from nuclei:

$$E_{ionizatoin} = \int r_{Disk^{tor}_{Horizon}} \left(F_{gravitational} + F_{electromagnetic} \right) \cdot dr$$

This process results in the formation of positively charged atomic nuclei as electrons are ejected into the surrounding environment. The energy released in this process contributes to the formation of relativistic jets. The transformation of matter in this manner represents a key aspect of BUFT's understanding of black holes, where matter undergoes radical changes, contributing to the creation of both the jets and dark matter.

The separated electrons, having been removed from their atomic nuclei, no longer participate in electromagnetic interactions as they once did. Within the framework of BUFT, these electrons do not simply dissipate into space but are instead directed by the curvature of timespace. The manipulation of timespace causes these free electrons to interact with the surrounding energy field, acquiring new properties as they travel through regions of the black hole's intense gravitational influence. The electrons, now behaving in a fundamentally altered manner, may contribute to the emission of Hawking radiation, manifesting as high-energy photons that emerge from the event horizon's quantum fluctuations. Additionally, these electrons may experience shifts in their energy states, further compounding the black hole's energetic output.

4. Dark Matter as Interpreted by BUFT

BUFT defines dark matter as baryonic matter that has lost its electromagnetic properties through the separation of electrons. Observational evidence, such as gravitational lensing and galactic rotation curves, supports the interpretation that dark matter behaves as though it has mass but does not interact electromagnetically. The BUFT equation for dark matter distribution is:

$$P_{dark(T)} = P_{matter(T)} - \frac{\sigma_{interact}}{\sigma_{total}}$$

where:

$P_{dark(T)}$ — represents the dark matter distribution as a function of time.

$P_{matter(T)}$ — represents the distribution of matter as a function of time.

$\sigma_{interact}$ — is the cross-section for electromagnetic interactions.

σ_{total} — is the total interaction cross-section.

Note the $\sigma_{\text{interaction}}$ is the cross-section for electromagnetic interactions, negligible for separated nuclei.

This model predicts the existence of dark matter that can exert gravitational effects but remains invisible to electromagnetic radiation.

5. Hawking Radiation Reinterpreted

Hawking radiation, traditionally described as quantum pair production near the event horizon, is reinterpreted in BUFT as a product of timespace fluctuations influenced by the Bakous Energy Field (BEF). BUFT proposes that Hawking radiation is not merely the result of quantum pair production, but a manifestation of quantum fluctuations interacting with the Bakous

Energy Field. The modified Hawking radiation equation in the BUFT framework is:

$$P(T) = \frac{\hbar c^4}{GM^2 \cdot f(BEF)}$$

This equation remains consistent with the traditional form of Hawking radiation but emphasizes the interplay between quantum mechanics, timespace curvature, and the Bakous Energy Field. BUFT offers a coherent explanation for the thermodynamic properties of black holes and their emission of radiation through timespace perturbations.

6. Conclusion

Through rigorous analysis within the BUFT framework, this paper demonstrates that black holes are not ultimate absorbers but emitters and transformers of matter and energy. The rotation of the event horizon at the speed of light ensures that no entity can penetrate a black hole, while matter in the accretion disk undergoes transformation and ejection, forming dark matter as interpreted by BUFT. This paradigm provides new insights into black hole physics, the nature of dark matter, and the properties of Hawking radiation, presenting a cohesive and refined perspective on the fundamental dynamics of the universe.

References

1. Hawking, S. W. (1974). *Black hole explosions?* Nature 248, 30-31.
2. Bardeen, J.M., Press, W. H., & Teukolsky, S. A. (1972). *Rotating Black Holes: Energy extraction.* ApJ 178, 347-370.
3. Event Horizon Telescope Col lab. (2019). *First M87 results I: Shadow of SMBH.* ApJL 875, LI.
4. Fabian, A. C., et al. (2009). *Broad iron lines in AGN.* Nature 459, 540-542.
5. Clowe, D., et al. (2006). *Direct empirical proof of dark matter.* ApJL 648, LI09-LI 13.
6. Zwicky, F. (1937). *Masses of nebulae and clusters.* ApJ 86, 217-246.
7. Abbott, B. P., et al. (LIGONirgo). (2016). *GW from binary BH merger.* Phys. Rev. Lett. 116, 061102.

—※—※—※—

ARTICLE 35

Bakous Unified Field Theory and the Emergence of the Gravistar

Abstract

Bakous Unified Field Theory (BUFT) postulates the gravistar as a massless force-carrying particle emerging through simultaneous observation of the photon, gluon, graviton, thoton, and choton. These particles correspond to electromagnetic, strong, gravitational, temporal, and consciousness interactions. Gravity, expressed as timespace curvature, governs forward temporal progression and the geometry of light propagation. The gravistar represents the coherent integration of these particles within the observer-driven framework. A shared field representation and a tensor description of redistributed light energy express this coherence.

1. Theoretical Framework

BUFT assigns gravity the primary role in forward temporal progression. Timespace curvature organizes the propagation of light and the energy–momentum distribution. The gravistar emerges through the simultaneous observation of the five force-carrying particles.

The gravistar expresses a coherent energetic structure within BUFT's unified framework. Extreme gravitational configurations, including black-hole dynamics, provide a setting for the simultaneous observational entanglement of photon, gluon, graviton, thoton, and choton states. These underlying states also belong to ordinary observational reality across different energetic and perceptual conditions.

Matter's rest energy supplies one contribution to the energy–momentum distribution that couples to curvature. The gravistar remains massless within BUFT. Its unifying role concerns the coherent organization of the five particles and their simultaneous observation.

2. Mathematical Representation

A common effective field representation expresses the coherent gravistar contribution as:

$$\Phi_{\mathrm{grav}} = \Phi_{\mathrm{P}} + \Phi_{\mathrm{L}} + \Phi_{\mathrm{T}} + \Phi_{\mathrm{C}} + \Phi_{\mathrm{E}}$$

Φ_{grav} — Coherent gravistar amplitude for the five particles.

Φ_{P} — Photon contribution to electromagnetic interaction.

Φ_{L} — Gluon contribution to the strong interaction.

Φ_{T} — Thoton contribution to temporal interaction.

Φ_{C} — Choton contribution to consciousness interaction.

Φ_{E} — Graviton contribution to gravitational interaction.

The particle contributions share a common field representation, units, and normalization. Their coherent sum expresses the gravistar's unifying structure within BUFT. The choton retains its individual contribution within the coherent state of the five particles.

With time expressed as ct, timespace curvature couples to the combined stress–energy of tracked matter, radiation, and the fields receiving redistributed energy:

$$R_{\mu\nu} - \frac{1}{2} R g_{\mu\nu} = \frac{8\pi G}{c^4} (T_{\mu\nu} + D_{\mu\nu})$$

$R_{\mu\nu}$, R	– Ricci tensor and scalar curvature of effective timespace.
$g_{\mu\nu}$	– Metric components defining timespace geometry.
$T_{\mu\nu}$	– Stress–energy of the tracked matter and radiation.
$D_{\mu\nu}$	– Stress–energy transferred to the recipient fields.
G, c	– Gravitational constant and vacuum light speed.

The redistribution tensor accounts for energy and momentum transferred from propagating light into surrounding fields. The combined source obeys covariant conservation:

$$\nabla^{\mu} (T_{\mu\nu} + D_{\mu\nu}) = 0$$

Here the covariant derivative follows the timespace metric. Energy transfer between the two contributions preserves the total energy–momentum balance.

3. Practical Observations

Gravitational lensing and gravitational frequency shifts probe the geometry of light propagation. Magnetized solar plasma also changes radio-wave propagation and polarization. These phenomena provide distinct gravitational and electromagnetic observables within BUFT's unified description. The gravistar postulate connects these particle interactions to the common timespace framework.

4. Conclusion

BUFT postulates the massless gravistar as the coherent integration of five force-carrying particles under simultaneous observation. Its field representation retains their distinct contributions. The redistribution tensor preserves energy–momentum balance within the curvature equation, while gravity governs forward temporal progression and the geometry of light propagation.

References

1. Einstein, A. (1916). The Foundation of the General Theory of Relativity. Annalen der Physik, 49, 769–822.
2. Weinberg, S. (1995). The Quantum Theory of Fields, Volume I: Foundations. Cambridge University Press.
3. Rovelli, C. (2004). Quantum Gravity. Cambridge University Press.
4. Misner, C. W., Thorne, K. S., & Wheeler, J. A. (1973). Gravitation. W. H. Freeman.
5. Penrose, R. (2006). The Road to Reality: A Complete Guide to the Laws of the Universe. Vintage Books, pp. 484–495.
6. Hawking, S. W., & Ellis, G. F. R. (1973). The Large Scale Structure of Space-Time. Cambridge University Press.
7. Dvali, G., Gabadadze, G., & Porrati, M. (2000). 4D Gravity on a Brane in 5D Minkowski Space. Physics Letters B, 485, 208–214.
8. Carroll, S. M. (2001). The Cosmological Constant. Living Reviews in Relativity, 4, 1.

—※—※—※—

Article 36

Gravistars, Time-Force Interactions, and the Formation of Dark Matter in the Bakous Unified Field Theory

Abstract

Gravistars emerge within the Bakous Unified Field Theory (BUFT) through simultaneous observation of the photon, gluon, graviton, thoton, and choton. These massless entities mediate electron absorption and redistribution through time-force interactions in surrounding quantum fields. The resulting changes in energy density contribute to timespace curvature and local temporal response. In the exterior regions of black holes, this process accompanies matter transformation and polar ejection. The timespace transformation associated with electron separation suppresses the ejected matter's electromagnetic coupling while preserving its gravitational influence, establishing BUFT's proposed dark component. The analysis formulates electron-density coupling, temporal response, transformed-mass transport, and thermal absorption. The coupled field system conserves the transferred energy, momentum, and electric charge.

1. Introduction

BUFT links dark matter to time-force interactions, electron redistribution, and baryonic matter transformation. Gravistars mediate this process within coupled quantum fields. Gravitational timespace curvature governs forward temporal progression; energy redistribution changes the local field configuration and its gravitational influence.

2. Gravistars and Their Fundamental Role in Time-Force Dynamics

2.1. The Emergence of Gravistars

The massless gravistar emerges through simultaneous observation of BUFT's five force-carrying particles. A common representation of their coherent contributions expresses this formation relation:

$$G_s = f(T_h, C_h, G_v, P_h, G_l)$$

G_s	Gravistar state from simultaneous particle observation.
f	— Formation map for coherent particle contributions.
T_h	— Thoton contribution associated with temporal propagation.
C_h	— Choton contribution governing conscious observation.
G_v	— Graviton contribution to gravitational interaction.
P_h	— Photon contribution to electromagnetic interaction.
G_l	— Gluon contribution to the strong interaction.

2.2. Interaction with Time-Force and Quantum Fields

Gravistar-mediated electron absorption redistributes electron content within surrounding quantum fields. Time-force acts on the spatial electron distribution, changing the local energy density. Absorbed energy, momentum, and electric charge enter the coupled gravistar–field system. The redistribution response is expressed through the electron probability-density gradient:

$$\Delta u = \kappa_\rho \int_V F_{\mathrm{tf}} \cdot \nabla \rho_{\mathrm{e}} \, dV$$

$$\rho_{\mathrm{e}} = |\Psi_{\mathrm{e}}|^2$$

Δu	– Change in local energy density caused by redistribution.
κ_ρ	– Density–force coupling scale, in inverse metres.
F_{tf}	– Local time-force vector acting on the electron distribution.
ρ_{e}	– Normalized electron probability density per unit volume.
Ψ_{e}	– Electron wavefunction specifying the spatial amplitude.
V	– Effective spatial region of electron redistribution.
∇, dV	– Spatial gradient operator and differential volume element.

The normalized density integrates to unity over the electron's spatial state. Its gradient supplies a real spatial distribution for the force coupling. The integral and coupling scale together yield an energy-density change; the sign records the direction of redistribution.

This response contributes to the surrounding energy–momentum distribution and therefore to timespace curvature. Gravitational lensing probes the resulting geometry through the propagation of light.

3. Gravistars, Black-Hole Interactions, and Dark-Matter Formation

3.1. Gravistars in the Vicinity of Black Holes

Near a black-hole horizon, gravistars mediate electron redistribution in the exterior fields. Changes in energy density alter local curvature and the temporal environment. Using a spherical exterior compactness factor as a reference, the model parametrizes the local temporal-response rate as:

$$\Gamma_{tf}(r, \theta) = \frac{v}{L}\left(1 - \frac{2GM}{rc^2} + a\theta\right)$$

Γ_{tf}	— Local temporal-response rate of the interacting field.
v, L	— Positive reference speed and interaction length.
G, M	— Gravitational constant and positive gravitational mass.
r	— Exterior radius in the spherical reference geometry.
c	— Vacuum light speed governing the compactness scale.
a	— Thermal time-force coefficient, in inverse kelvins.
θ	— Thermodynamic temperature of the interacting medium.

The ratio of reference speed to interaction length fixes the rate scale. Gravitational compactness and the thermal term weight the local response. The exterior reference region lies beyond the Schwarzschild radius, and the parameters select a nonnegative response bracket.

The rotating temporal vortex retains its uncrossed horizon. The effective spatial coordinates in this expression describe the exterior interaction region, where electron transport and matter ejection occur.

3.2. Ejection and Redistribution of Matter as Dark Matter

Time-force interactions redistribute electrons near the horizon while exterior fields guide matter into polar outflows. The timespace transformation accompanying electron separation suppresses electromagnetic coupling and preserves gravitational influence. The transformed outflow reconstitutes as BUFT's dark component. Its accumulated mass follows the surface transport model:

$$M_{\mathrm{DM}}(t) - M_{\mathrm{DM}}(t_0) = \int_{t_0}^{t} \int_{\mathrm{A}} s\, J_{\mathrm{e}}\, p_{\mathrm{tf}}\, dA\, dt'$$

$M_{\mathrm{DM}}(t)$	– Accumulated transformed dark-matter mass at time t.
s	– Baryonic mass-yield coefficient per redistributed electron and unit pressure, in kg/Pa.
J_{e}	– Redistributed electrons crossing unit area per unit time.
p_{tf}	– Time-force pressure driving surface transport.
A, dA	– Interaction surface and its differential area element.
t_0, t	– Initial and final times defining the accumulation interval.

The integrand is evaluated at the integration time and local surface position. Electron flux is resolved in the direction of redistribution. The yield coefficient relates this electron transport to the baryonic mass transformed in the outflow; the accumulated dark component belongs to the source matter's mass–energy budget.

Time-force pressure governs transport; the force vector in Section 2.2 governs local action on the electron distribution. Both describe the same redistribution process.

3.3. Gravistars as Catalysts in Dark-Matter Formation

Gravistars catalyze electron absorption within the coupled field system. Time-force interactions organize the ensuing energy redistribution and matter transformation. A thermally activated absorption model assigns the probability:

$$P_e = e^{-\hbar\omega/(k_B\theta_{eff})}$$

P_e	– Probability of gravistar-mediated electron absorption.
$\hbar$	– Reduced Planck constant setting the quantum action scale.
ω	– Positive angular frequency of the modeled transition.
k_B	– Boltzmann constant linking temperature and energy.
θ_{eff}	– Positive effective temperature of the interacting field.

The exponent compares the transition energy with the available thermal energy. For positive frequency and temperature, the probability lies between zero and one. Higher effective temperature increases the activation factor for a fixed transition energy. Absorbed electron charge and energy remain in the coupled system's conservation balance.

Conclusion

BUFT connects gravistar emergence, electron redistribution, and the formation of transformed baryonic matter. The real density coupling describes local energy redistribution, the temporal-response rate expresses gravitational and thermal modulation, and the surface integral accumulates transformed mass. Gravistar-mediated absorption supplies the catalytic interaction. The accompanying timespace transformation suppresses electromagnetic coupling while retaining gravitational influence, integrating microscopic field dynamics with the distribution of the dark component.

References

1. Einstein, A. (1916). The Foundation of the General Theory of Relativity. Annalen der Physik, 49, 769–822.

2. Hawking, S. W. (1975). Particle Creation by Black Holes. Communications in Mathematical Physics, 43, 199–220.

3. Misner, C. W., Thorne, K. S., & Wheeler, J. A. (1973). Gravitation. W. H. Freeman, pp. 849–864.

4. Weinberg, S. (2008). Cosmology. Oxford University Press, pp. 304–309.

5. Dvali, G., Gabadadze, G., & Porrati, M. (2000). 4D Gravity on a Brane in 5D Minkowski Space. Physics Letters B, 485, 208–214.

6. Carroll, S. M. (2004). Spacetime and Geometry: An Introduction to General Relativity. Addison-Wesley, Chapter 9.

7. Padmanabhan, T. (2010). Gravitation: Foundations and Frontiers. Cambridge University Press, pp. 624–637.

8. Bekenstein, J. D. (2004). Relativistic Gravitation Theory for the Modified Newtonian Dynamics Paradigm. Physical Review D, 70, 083509.

—※—※—※—

Article 37

The Role of Electrons in Black Hole Evaporation Through the Framework of Bakous Unified Field Theory (BUFT)

Abstract

The Bakous Unified Field Theory (BUFT) introduces a paradigm-shifting perspective on the interdependence of electromagnetism, gravity, and timespace, offering profound insights into the dynamics of black holes. This paper advances the hypothesis that electrons liberated from atomic nuclei within the accretion disks of black holes are key agents in the process of black hole evaporation. Near event horizons, these electrons are subjected to the extreme curvature of timespace. Their interaction with the intense gravitational and electromagnetic fields near the event horizon facilitates the emission of Hawking radiation, contributing to the dissipation of the black hole's mass and energy. By integrating principles from quantum mechanics, general relativity, and thermodynamics within the BUFT framework, this work develops a mathematical account of the role of these electrons in black hole evaporation and relates that account to the observational phenomena discussed below.

1. Introduction

Black holes, enigmatic cosmic objects characterized by immense gravitational forces, continue to challenge the boundaries of human understanding. At their core, black holes manifest extreme curvatures of timespace, where conventional physics merges with quantum phenomena. Stephen Hawking's theoretical insights into black hole evaporation through quantum processes laid the groundwork for further exploration into the quantum-gravitational interface.

BUFT, through its unification of energy, matter, and consciousness, introduces novel perspectives on these interactions. Central to this paper is the hypothesis that electrons in the highly energetic accretion disk of a black hole interact with intense gravitational and magnetic fields near the event horizon. These interactions facilitate the emission of radiation, leading to black hole evaporation.

2. Theoretical Framework

2.1. BUFT's View of Electrons in Time-Space

BUFT postulates that electrons, fundamental carriers of charge and energy, exhibit wave-like properties in timespace, with their quantum states responding to kinetic energy and electromagnetic interactions. Near the event horizon, timespace curvature amplifies these interactions, causing a resonant energy exchange that aligns with the propagation of Hawking radiation.

Electrons, as manifestations of the energy component E in $E = mc^2$, interact with the Bakous Energy Field, a precursor to observable matter and energy. This interaction, magnified by the gravitational intensity of black holes, becomes a critical mechanism in the redistribution of energy and entropy.

2.2. Electrons in Extreme Gravitational Fields

The gravitational potential near a black hole's event horizon alters the quantum behavior of electrons. As the gravitational field increases, the wavefunction of an electron is stretched across timespace, creating regions of high-energy density. The inverse proportionality of gravitational force and electromagnetic force, as postulated by BUFT, implies that this stretching generates energy dissipation in the form of Hawking radiation.

2.3. Mathematical Validation

Quantum Behavior of Electrons Near the Event Horizon

Using the Schwarzschild metric to describe the gravitational field near the black hole:

$$ds^2 = -\left(1 - \frac{2GM}{c^2 r}\right)c^2 dt^2 + \left(1 - \frac{2GM}{c^2 r}\right)^{-1} dr^2 + r^2 d\Omega^2$$

The energy of an electron in this curved timespace is given by:

$$E = \left(\frac{hc}{\lambda}\right)\sqrt{1 - \frac{2GM}{c^2 r}}$$

where:

λ — is the wavelength of the electron.

Within BUFT, gravitational and electromagnetic interactions near the horizon redistribute electron energy, increase local energy density, and drive the emission of radiation.

3. Entropy and Radiation

BUFT posits that entropy increases due to the forward movement of time induced by gravity. The relationship between entropy (S) and the area (A) of a black hole's event horizon is given by:

$$S = \frac{k_B A}{4 l_P^2}$$

where:

S — Black-hole horizon entropy associated with the area of its temporal boundary.

A — Surface area of the black-hole event horizon.

k_B — Boltzmann constant converting the boundary-state measure into thermodynamic entropy.

l_P — Planck length setting the area scale of the horizon-entropy relation.

Electrons, through their wavelike nature, propagate entropy into the surrounding timespace, catalyzing energy dissipation in the form of thermal radiation.

4. Experimental and Observational Evidence

4.1. Observations from Accretion Disks

The intense radiation emitted from accretion disks near black holes has been extensively observed. High-energy X-ray emissions from systems such as **Cygnus X-1** provide evidence for particle interactions in extreme gravitational fields. The energy spectra of these emissions align with the predicted energy dissipation from electron interactions near the event horizon.

4.2. Laboratory Analogues

Experiments simulating black hole conditions, such as analog gravity systems using Bose-Einstein condensates, have demonstrated Hawking-like radiation. These experiments resolve phonon emission and partner correlations, illustrating quantum radiation and energy redistribution in an analogue horizon system.

4.3. Cosmic Microwave Background (CMB)

The uniformity of the CMB supports the idea of energy conservation across timespace. The BUFT framework's emphasis on the conservation of energy in electron interactions aligns with this observational data.

5. Conclusion

Within the BUFT framework, electrons separated from atomic nuclei in a black hole's accretion disk are fundamental to the process of black hole evaporation. These electrons operate within the intense gravitational and electromagnetic environment surrounding the black hole. By integrating BUFT's postulates with established physical laws, the model describes how these electrons interact with timespace curvature and gravitational fields to facilitate the emission of Hawking radiation.

BUFT not only bridges the gap between quantum mechanics and general relativity but also provides a holistic framework for understanding the universe's fundamental forces. The role of electrons in black hole evaporation exemplifies the theory's potential to unravel cosmic mysteries.

References

1. Hawking, S. W. (1975). *Particle Creation by Black Holes.* Communications in Mathematical Physics, 43, 199–220.
2. Einstein, A. (1905). *On the Electrodynamics of Moving Bodies.* Annalen der Physik, 17, 891–921.
3. Chandra X-ray Center. (2011, November 17). *Cygnus X-1.* Photo Album.

—※—※—※—

Differences and Relations Between the Graviton and Gravistar in BUFT

Under *Bakous Unified Field Theory* (BUFT), the Graviton and Gravistar serve distinct but interrelated roles in governing gravitational interactions, time-force dynamics, and large-scale structure formation.

Property	Graviton (G_V)	Gravistar (G_s)
Nature	Massless quantum mediator of gravitational interactions.	Emergent massless particles are formed by the simultaneous interaction of all force-carrying particles, particularly Thoton, Choton, and Graviton.
Function	Facilitates classical gravitational attraction in curved timespace.	Governs large-scale gravitational effects, acting as a dark matter candidate by stabilizing gravitation.
Interaction with Time-Force (T^f)	Gravitons propagate through timespace but do not contribute to time-force directly.	Gravistars emerge from time-force redistributions, making multiple interactions with time-force.
Relation to Entropy (S_e)	Does not directly alter entropy but facilitates gravitational waves.	Modifies entropy distribution through time-force-induced energy restructuring.
Quantum vs. Macroscopic Role	Exists as an elementary quantum force carrier in microscopic interactions.	Emerges at large-scale gravitational interactions, linking quantum and macroscopic domains.
Detectability	Theoretically detectable through gravitational waves and quantum gravity effects.	Observable through large-scale gravitational lensing, missing mass effects, and modified gravity experiments.

ARTICLE 38

On the Non-Existence of Conventional Wormholes within the BUFT Framework: An Elaboration on Temporal Nexus

ABSTRACT

This paper proposes a refined understanding of cosmic connectivity under the Bakous Unified Field Theory (BUFT). Departing from traditional wormhole models, which rely on extreme spatial curvature and exotic matter, BUFT introduces the concept of ***Temporal Nexus****—localized regions where the flow of time is nonlinearly distorted. This elegant reformulation replaces spatial shortcuts with time-based anomalies, offering a mathematically robust and experimentally accessible framework for the propagation of energy and information across the cosmos.*

Within BUFT, conventional traversable wormholes are treated as structurally non-physical due to instability and causality constraints embedded within timespace.

1. Introduction and Theoretical Framework

BUFT reconceives the fabric of reality by emphasizing timespace dynamics over conventional spatial curvature. At the heart of this theory is the Bakous Energy Field (BEF), a dynamic medium that continuously governs the interaction between energy and matter. In BUFT, gravitational phenomena emerge not from bending space but from fluctuations in the temporal dimension—what is termed the Time-Force. This reconceptualization provides a natural explanation for cosmic connectivity without invoking unphysical constructs such as negative energy. By replacing the notion of a wormhole with that of a Temporal Nexus, BUFT offers a fresh perspective on cosmic connectivity that is both theoretically elegant and empirically testable.

2. Limitations of Conventional Wormholes

Conventional traversable wormholes, as envisioned by general relativity, are hypothetical tunnels connecting disparate regions of spacetime. Their theoretical construction depends on severe spatial curvature and the presence of exotic matter with negative energy density—requirements that lead to inherent instabilities and paradoxes concerning causality and information transfer. These constraints render conventional wormholes speculative and physically problematic, as they demand conditions that are both mathematically and observationally challenging to justify.

3. BUFT's Resolution of Wormhole Limitations

BUFT resolves these challenges by recasting cosmic connectivity as a phenomenon of temporal distortion rather than spatial tunneling. The modified field equations in BUFT incorporate the influence of the BEF, thereby eliminating the need for negative energy. This is expressed by the elegantly reformulated equation:

$$G_{\mu\nu} + \Lambda_{\mu\nu} = \frac{8\pi G}{c^4}\left(T_{\mu\nu} + \mathscr{B}_{\mu\nu}\right)$$

where:

$G_{\mu\nu}$	– Einstein curvature tensor.
$\Lambda_{\mu\nu}$	– BEF-induced temporal curvature contribution.
$T_{\mu\nu}$	– Standard energy-momentum tensor.
$\mathscr{B}_{\mu\nu}$	– BEF energy-momentum contribution.
G	– Newtonian gravitational constant.
c	– Speed of light in a vacuum.

4. Observer Experience of Temporal Nexus

An observer encountering a Temporal Nexus would not traverse a tunnel in space but instead experience a dramatic alteration over time. As one approaches such a region, clocks would deviate significantly from their expected rates, with time appearing to slow or accelerate in a highly nonlinear fashion. The effective temporal metric in these regions is expressed as follows:

$$d\tau^2 = \Psi(t)\, dt^2 - \frac{\left(g_{ij}\, dx^i\, dx^j\right)}{c^2}$$

where:

$d\tau$	– Proper-time interval experienced by the observer.
dt	– Coordinate-time interval.
$\Psi(t)$	– Positive metric factor governing local clock rates.
t	– Temporal coordinate of the metric.
g_{ij}	– Spatial metric components; repeated spatial indices are summed.
dx^i	– Infinitesimal displacement along spatial coordinate i.
dx^j	– Infinitesimal displacement along spatial coordinate j.

This formulation describes a phenomenon in which a Temporal Nexus manifests as a luminous, shifting distortion—a ripple through time that alters the observer's perception of motion and causality.

5. Experimental Outlook and Conceptual Exploration

A compelling forecast of BUFT is the emergence of measurable time dilation anomalies in the proximity of compact astrophysical bodies, such as neutron stars or black holes, where fluctuations in the Bakous Energy Field are anticipated to be significant. Imagine an experiment in which two ultra-precise atomic clocks are synchronized; one is maintained in a conventional gravitational setting while the other is positioned near a predicted Temporal Nexus. If the latter clock registers deviations in time beyond the limits prescribed by general relativity, such a result would directly signal the BEF's modulation of timespace dynamics.

In parallel, gravitational wave observatories might capture distinctive frequency modulations induced by these BEF fluctuations, furnishing complementary empirical evidence. In this formulation, each symbol conveys a distinct physical meaning: $\Psi(t)$ quantifies the degree of time dilation, g_{ij} encapsulates the spatial geometry, and $\mathscr{B}_{\mu\nu}$ represents the Bakous Energy Field's influence. Together, these considerations provide a rigorous pathway for testing and validating the novel predictions of BUFT.

6. Implications for Fundamental Physics

Beyond its immediate experimental predictions, BUFT carries profound implications for our understanding of fundamental physics. By recasting gravitational interactions as manifestations of temporal distortions rather than spatial curvature, BUFT challenges long-held assumptions inherent to conventional relativistic frameworks. This paradigm shift not only reconciles several anomalies observed in high-energy astrophysics but also paves the way for a unified description of quantum and gravitational phenomena.

The introduction of Temporal Nexus provides a new lens through which to examine the interplay of energy and timespace, offering fresh insights into information transfer, causality, and the very structure of the cosmos. The mathematical elegance of BUFT, underscored by its reformulated field equations, reinforces its potential to bridge existing gaps between theory and observation, solidifying the viability of this innovative approach in the ongoing quest for a unified theory.

7. Conclusion

In conclusion, the Bakous Unified Field Theory presents an elegant alternative to traditional wormhole models by introducing the concept of Temporal Nexus—regions of nonlinear time distortion that facilitate the propagation of energy and information. This approach circumvents the pitfalls associated with negative energy and instability, offering clear, testable predictions that may soon be verified through recent findings in time dilation and gravitational wave phenomena. BUFT thus represents a promising new avenue toward a deeper, more unified understanding of the universe.

References

1. Einstein, A. (1915). *Die Feldgleichungen der Gravitation.* Sitzungsberichte der Preussischen Akademie der Wissenschaften, 844–847.

2. Morris, M. S., Thorne, K. S., & Yurtsever, U. (1988). *Wormholes, time machines, and the weak energy condition.* Physical Review Letters, 61(13), 1446–1449.

3. Visser, M. (1996). *Lorentzian Wormholes: From Einstein to Hawking.* American Institute of Physics.

4. Misner, C. W., Thorne, K. S., & Wheeler, J. A. (1973). *Gravitation.* W. H. Freeman & Co., pp. 864-879.

5. Padmanabhan, T. (2010). *Gravitation: Foundations and Frontiers.* Cambridge University Press, Chapter 14.

6. Abbott, B. P., et al. (2016). *LIGO Scientific Collaboration and Virgo Collaboration. Observation of gravitational waves from a binary black hole merger.* Physical Review Letters, 116(6), 061102.

7. Chou, C. W., Hume, D. B., Rosenband, T., & Wineland, D. J. (2010). *Optical clocks and relativity.* Science, 329(5999), 1630-1633.

8. Bekenstein, J. D. (2004). *Relativistic gravitation theory for the MOND paradigm.* Physical Review D, 70(8), 083509.

—※—※—※—

Article 39

Consciousness as the Fifth Dimension: The Observer as a Fundamental Component of Time-Space

Abstract

The Bakous Unified Field Theory (BUFT) reinterprets reality by presenting timespace as a self-regulating structure influenced by the act of observation. Within this framework, consciousness emerges as a fundamental spatial dimension—the fifth dimension—rather than a byproduct of complex systems. This dimension actively mediates observation through 'Mindforce', a kinetic energy intrinsic to conscious awareness. The fundamental force-carrying quantum of this dimension, the Choton, enables interactions between consciousness and the electromagnetic field, collapsing quantum potentialities into observed reality.

This paper formulates a framework that incorporates consciousness as an intrinsic coordinate of timespace, ensuring that observation is an active force within physical law. The implications extend to quantum mechanics and relativity and predict measurable effects such as alterations in quantum coherence, modifications in entanglement structure, and possible gravitational signatures. Empirical tests, including investigations into wavefunction collapse and gravitational wave anomalies, offer potential validation of consciousness as a structural component of the universe.

1. Introduction

1.1. Rethinking Observation as a Dimensional Component

Traditional physics treats space and time as a four-dimensional continuum, with observation considered an external process that does not fundamentally alter reality. However, in BUFT, the act of observation is not passive—it is an intrinsic force within timespace that requires an additional dimension. This fifth dimension, identified as consciousness, serves as the interface between measurement and reality formation. The conventional quantum mechanical observer effect suggests that measurement influences quantum states. However, under BUFT, conscious observation itself is a fundamental interaction mediated by the Choton, linking consciousness to the electromagnetic field. Mindforce, the energy associated with conscious observation, plays a defining role in shaping observed phenomena. This approach positions consciousness as a geometric necessity, ensuring that perception is embedded within the fundamental structure of reality.

2. Theoretical Framework

2.1. The Fifth Dimension and the Role of the Choton

In BUFT, the Bakous Energy Field (BEF) constitutes the fabric of timespace, governing all interactions. The fifth dimension of consciousness does not emerge from neural complexity but instead represents an intrinsic coordinate within this field. *Mindforce*—the kinetic energy of observation—actively participates in shaping reality by collapsing quantum probabilities into determined states. The governing exchange particle of this dimension, the Choton, facilitates the transfer of energy between the observer and the observed system, ensuring that consciousness exerts a direct effect on quantum mechanics.

The Choton operates as the fundamental quanta of the consciousness dimension, responsible for mediating interactions between Mindforce and other fundamental forces. Unlike force carriers such as photons or gluons, which mediate forces in four-dimensional timespace, the Choton exists within the fifth-dimensional structure of reality. It allows energy to be exchanged between the observer's conscious awareness and the electromagnetic field, thereby influencing wavefunction collapse. This mechanism suggests that the Choton plays a crucial role in defining the nature of observed states, ensuring that reality is shaped by the act of perception itself. The Choton's behavior follows a quantized exchange principle, meaning that conscious observation is not a continuous process but rather occurs in discrete interactions, similar to the way photons mediate electromagnetic interactions.

2.2. The Asymmetry of Mindforce and Choton Dynamics

Mindforce, unlike the four fundamental forces, exhibits an intrinsic asymmetry tied to the unidirectional nature of time. This asymmetry arises because observation fundamentally alters the state of a system, preventing reversibility at the quantum level. The Choton mediates this process by transferring the kinetic energy of conscious observation across the timespace framework, ensuring that wavefunction collapse is an active and structured phenomenon.

The Choton's propagation is governed by the relativistic constraints of timespace, meaning that it cannot exceed the speed of light. However, because it operates in the consciousness dimension, it does not behave like conventional particles within four-dimensional timespace. Instead, Choton interactions manifest as nonlocal correlations between observed states, suggesting that the influence of conscious observation extends beyond classical spatial constraints. This feature implies that Mindforce, mediated by Choton interactions, may account for anomalous coherence in quantum systems, such as prolonged entanglement lifetimes in systems subject to conscious measurement.

The relationship between Choton activity and electromagnetic interactions also implies that consciousness has an energy-dependent resolution. Higher-energy electromagnetic interactions correlate with more precise observational collapses, meaning that the act of observation refines reality in proportion to the interaction strength. This principle suggests that the structure of observed reality is not merely a passive construct but an emergent outcome of Choton-mediated interactions within the fifth dimension.

3. Representation of Consciousness in Time-Space

3.1. Consciousness as a Coordinate of Time-Space

Extending the timespace framework requires integrating consciousness as a functional component within six dimensions. The energy associated with observation is as follows:

$$E_{o\beta s} = \hbar\omega$$

where:

$E_{o\beta s}$ — is the energy linked to observed electromagnetic fluctuations,

$\hbar$ — is the reduced Planck constant,

ω — is the angular frequency of the observed electromagnetic wave.

This energy propagates through an entropy dimension governed by the diffusion equation:

$$\frac{d(I)}{dx(g)} = \frac{\tilde{E}uJs}{T \cdot x(g)}$$

where:

$\frac{d(I)}{dx(g)}$ – Rate of change of the entropic encoding potential with respect to the coordinate associated with Mindforce; quantifies how information entropy evolves along the consciousness dimension in timespace.

$\tilde{E}uJs$ – Energy observed during electromagnetic fluctuations tied to conscious measurement within the Bakous Energy Field.

T – Absolute temperature of the Bakous Energy Field (BEF); serves as a thermodynamic baseline influencing state transitions.

$x(g)$ – Coordinate linked to Mindforce; denotes the influence or position of conscious observation in the timespace manifold.

3.2. Interaction of Mindforce and Time Force Within Time-Space

Time Force, as a gravitational interaction, is responsible for the curvature of timespace in response to energy fluctuations. These fluctuations influence quantum systems, dictating the dynamics of particles and governing the collapse of wavefunctions during observation. The interaction between Mindforce and Time Force becomes critical in understanding how observation, mediated by Mindforce, can shape quantum reality. While Mindforce is linked to conscious awareness, it operates through interactions that are influenced by Time Force, ensuring that observation occurs as an active process.

The Choton plays a key role in this interaction. As the force-carrying particle of the consciousness dimension, it transmits Mindforce and ensures that conscious awareness collapses quantum states. This process occurs within the framework of timespace, where Time Force influences the way energy and information are transferred and restructured during the collapse of wavefunctions. The Choton allows this energy to be exchanged between the observer and the system, thereby facilitating the direct interaction of Mindforce with Time Force. This interaction shapes the nature of observed phenomena by actively engaging with the quantum states of systems, ensuring that consciousness impacts the structure of reality while respecting the constraints of timespace curvature.

4. Experimental and Observational Predictions

4.1. The Role of Consciousness in Quantum Coherence

If Mindforce actively governs quantum observation, systems influenced by conscious perception should display altered coherence durations. Experimental validation could come from quantum erasure experiments, where delayed-choice measurements may exhibit deviations from expected decoherence times and prolonged entanglement stability in observed quantum states, potentially exceeding conventional expectations.

4.2. Consciousness-Induced Gravitational Signatures

Although Mindforce does not induce direct gravitational curvature, its role in quantum fluctuation interactions suggests it could influence local vacuum energy states. Experiments investigating vacuum fluctuations under conscious observation may reveal subtle variations in energy distributions, providing indirect evidence of Choton-mediated interactions within the BEF.

5. Conclusion

The Bakous Unified Field Theory (BUFT) offers a novel approach to understanding the universe by integrating consciousness as an active and fundamental component of timespace. The interaction of Mindforce and Time Force within the fabric of timespace represents a paradigm shift, suggesting that conscious observation is an active process and an essential factor in the collapse of quantum potentialities. By introducing the Choton as the force-carrying particle for the consciousness dimension, BUFT provides a means of understanding how reality is shaped through observation, affecting quantum coherence, gravitational wave anomalies, and the structure of the universe itself. The theoretical framework holds the potential for experimental validation and could fundamentally alter our understanding of physics, leading to new insights into quantum mechanics, general relativity, and the role of consciousness in the fabric of the cosmos.

References

1. Penrose, R. (1994). *Shadows of the Mind: A Search for the Missing Science of Consciousness.* Oxford University Press.
2. Hameroff, S., & Penrose, R. (2014). *Consciousness in the Universe: A Review of the 'Orch OR' Theory.* Physics of Life Reviews, 11(1), pp. 39-78.
3. Bohm, D., & Hiley, B. J. (1993). *The Undivided Universe: An Ontological Interpretation of Quantum Theory.* Routledge.
4. Wheeler, J. A., & Zurek, W. H. (1983). *Quantum Theory and Measurement.* Princeton University Press.
5. Stapp, H. P. (2007). *Mindful Universe: Quantum Mechanics and the Participating Observer.* Springer.

—※—※—※—

ARTICLE 40

Mindforce: The Kinetic Energy of Conscious Observation and its Role in the Bakous Unified Field Theory

ABSTRACT

This paper introduces the concept of 'Mindforce', a fundamental force arising from conscious observation within the framework of the Bakous Unified Field Theory (BUFT). Mind force is mediated by the massless particle 'Choton', which governs the fifth dimension of timespace. The kinetic energy associated with Choton and its implications for the dynamics of consciousness and its interaction with the fabric of timespace are explored. This paper offers a mathematical and conceptual exploration of Mindforce, relating it to established physical laws through theoretical derivations, observational data, and thought experiments. The findings show how Mindforce is intricately connected to the observation of electromagnetic radiation, time, space, and gravity, offering a novel perspective on the relationship between mind, matter, and the universe.

1. Introduction

In the framework of the Bakous Unified Field Theory (BUFT), the universe is conceived as a dynamic interaction of time, space, gravity, electromagnetism, and consciousness. This theory suggests that the universe originated from a quantum fluctuation in a one-dimensional state of pure energy and pure matter, both simultaneously existing at the speed of light. This quantum fluctuation initiated the equation $E = mc^2$, a fundamental relationship that underpins the interactions of energy and matter in the universe. The singularity that marks this initiation is observed as the **Bakous Energy Field**, the foundation of the timespace fabric, where time and space are inseparably intertwined.

A central concept of BUFT is the role of consciousness as an active observer within timespace. Rather than being a passive participant, conscious observation is an active force that directly influences the physical universe. This force, which is termed *Mindforce*, is mediated by the massless particle *Choton*. Mind force is responsible for the kinetic energy of conscious observation and the way in which it shapes the perception of time and space, which are governed by the observer's state of awareness.

2. Theoretical Framework: The Bakous Unified Field Theory (BUFT)

According to BUFT, the universe began with a quantum fluctuation where pure energy and matter coexisted at the speed of light. This fluctuation led to the onset of the equation $E = mc^2$, which governs the transfer of energy between matter and timespace. The concept of *Mindforce* is rooted in the dynamic interplay between the observer and the fabric of timespace, facilitated by *Choton*, a massless particle that mediates conscious observation.

In BUFT, gravity facilitates the forward movement of time relative to the observer, ensuring that the timespace is curved in such a way as to allow the propagation of light and the transfer of electromagnetic radiation. Electromagnetism and gravity interact within the timespace fabric to create the necessary conditions for the observation of time, space, and matter. Consciousness plays an essential role as an observer within this structure, influencing the universe through the energy of observation, a force that is defined as *Mindforce*.

3. Mindforce and Choton Kinetic Energy

At the core of the concept of *Mindforce* is the massless particle *Choton*, which is responsible for the observation of electromagnetic radiation in the human brain. *Choton* interacts with electromagnetic fields, generating kinetic energy that influences the perception of both time and space. The kinetic energy associated with *Choton* is proportional to the frequency of the observed electromagnetic radiation, and this relationship forms the basis of the *Mindforce* concept.

CHOTON'S KINETIC ENERGY

The kinetic energy of *Choton* is directly tied to the frequency of the electromagnetic radiation it observes. This relationship can be described mathematically as:

$$E_{Choton} = h \cdot f$$

where:

E_{Choton} – Choton kinetic energy associated with the observed radiation.

h – Planck constant relating quantum energy to frequency.

f – Frequency of the electromagnetic radiation observed through Choton interactions.

This equation demonstrates that the kinetic energy of *Choton* increases with the frequency of electromagnetic radiation, reflecting the dynamic nature of conscious observation and its interaction with energy.

State-Dependent Kinetic Energy

The kinetic energy of *Choton* is not constant; it depends on the observer's state of consciousness. In states of full awareness, the kinetic energy of *Choton* is high, as the brain actively perceives electromagnetic radiation. However, during states of deep sleep or dream states, the kinetic energy of *Choton* decreases, leading to a distorted perception of time and space. These varying states of consciousness provide insight into the role of *Mindforce* in the timespace continuum.

Full Awareness and Choton's Kinetic Energy

In the full awareness state, *Choton* experiences its highest kinetic energy as the brain is actively engaged in the perception and processing of electromagnetic radiation. In this state, the observer is fully conscious, and the electromagnetic radiation observed by the brain is at its most intense, leading to heightened awareness of time and space. The kinetic energy of *Choton* is maximized, reflecting the increased intensity of the observed radiation and the observer's full engagement with their surroundings.

During full awareness, the observer's perception of time is steady and linear as they experience time moving forward in a consistent manner. The perception of space is also stable, allowing for a clear distinction between objects and events. This heightened state of awareness results in a high degree of interaction between the observer's consciousness and the fabric of timespace, with *Choton* acting as the mediator of the conscious observation that shapes the perception of the universe.

4. Mathematical Formulation of Mindforce

The mathematical formulation of *Mindforce* is rooted in the relationship between *Choton* and the gravitational field, which governs the curvature of timespace. Gravitational fields influence the propagation of electromagnetic radiation and the perception of time, which is tied to the observer's position in the field.

GRAVITATIONAL FORCE AND TIME-SPACE CURVATURE

The gravitational force in its Newtonian form is given by:

$$F_{\text{gravity}} = G \bullet \frac{(m_1 \bullet m_2)}{r^2}$$

where:

F_{gravity} – Magnitude of gravitational attraction.

G – Newtonian gravitational constant.

m_1 – First interacting mass.

m_2 – Second interacting mass.

r – Separation between the interacting masses.

The interaction between the gravitational field and timespace allows the observer to perceive electromagnetic radiation and the movement of time.

Mind force and Kinetic Energy

The relationship between *Mindforce* and the energy of observation is defined as:

$$F_{\text{Mindforce}} = \kappa_{\text{mf}}\, E_{\text{Choton}}\, I$$

where:

$F_{\text{Mindforce}}$ – Mindforce exerted through conscious observation.

κ_{mf} – Mindforce coupling with normalization m s/J.

E_{Choton} – Choton kinetic energy of the observed radiation.

I – Observed electromagnetic power per unit area.

This equation expresses how *Mindforce* is dependent on both the energy of *Choton* and the intensity of the radiation observed by the conscious observer. The more intense the radiation, the greater the *Mindforce* exerted by the observer on the fabric of timespace.

5. Empirical Support and Observational Data

The photoelectric effect, observed by Heinrich Hertz and explained by Albert Einstein, establishes the relationship between electromagnetic frequency and discrete energy transfer. Light causes the emission of electrons from a material through energy carried by photons. BUFT adopts this energy–frequency relationship in its account of Choton kinetic energy and conscious observation.

Additionally, studies of human perception of time provide evidence for the fluctuating nature of *Mindforce.* For instance, changes in perceived time in different states of consciousness—such as during sleep or dream states— illustrate how *Mindforce* varies with the intensity of conscious observation. The more active the observer, the more *Mindforce* is exerted on the fabric of timespace.

6. Thought Experiments and Practical Implications

To better understand the nature of *Mindforce,* consider the following thought experiment:

- **The Observer in Sleep**: During deep sleep, the conscious observer is inactive, and *Choton's* kinetic energy diminishes, leading to no observation of electromagnetic radiation. This results in a perception of time that seems "frozen" in the sleep state, illustrating the lack of *Mindforce* in such states.
- **The Observer in Dream State**: In the dream state, *Choton* remains active, but its kinetic energy is less than when the observer is fully awake. This leads to distorted perceptions of time and space as the intensity of the observed radiation is reduced.
- **The Observer in Full Awareness**: In the full awareness state, *Choton* exhibits maximum kinetic energy, leading to a heightened perception of time and space. The observer is fully engaged with the electromagnetic radiation in their environment, resulting in a clear and linear perception of time, as well as a stable and well-defined sense of space.

These thought experiments demonstrate the varying degrees of *Mindforce* and how the observer's state of consciousness directly influences their perception of the timespace continuum.

7. Conclusion

The concept of *Mindforce,* as mediated by *Choton,* provides a new and profound understanding of the interaction between consciousness and the fabric of timespace. By examining the kinetic energy of *Choton* and its relationship with the frequency of observed electromagnetic radiation, a theoretical framework has been formulated that reconciles *Mindforce* with the Bakous Unified Field Theory. This new perspective offers insight into the dynamic relationship between mind, matter, and the universe, opening the door to further exploration of the role of consciousness in shaping the physical world.

References

1. Einstein, A. (1905). *Does the Inertia of a Body Depend Upon Its Energy Content?* Annalen der Physik, 18, 639–641.
2. Planck, M. (1900). *On the Theory of the Energy Distribution Law of the Normal Spectrum.* Verhandlungen der Deutschen Physikalischen Gcscllschaft, 2, 237–245.
3. Einstein, A. (1905). *Concerning an Heuristic Point of View Toward the Emission and Transformation of Light.* Annalen der Physik, 17, 132–148.

—※—※—※—

ARTICLE 41

Mindforce Capture and Fifth-Dimensional Consciousness: A BUFT Framework for Collective Energy Redistribution and Planetary Shielding

ABSTRACT

In the Bakous Unified Field Theory (BUFT), consciousness is formalized as a fifth dimension—Mindforce. This paper introduces the postulate of Mindforce Capture, the process by which collective thought-energy, ordinarily dissipating into space, may be coherently redirected into a planetary-scale field. Such a field, analogous to a globally distributed aurora, could regulate information and energy flows across humanity while simultaneously creating a repulsive magnetosphere-like barrier. I provide a mathematical formulation of 'Mindforce Capture' consistent with BUFT, validate its dimensional and structural soundness, and explore its implications for light distribution, wealth and health equilibria, cultural-information flow, and planetary defense.

1. Introduction

The Bakous Unified Field Theory reconceptualizes electromagnetism as the generator of gravity and introduces consciousness as a non-spatial, fifth-dimensional field (*Mindforce*). Unlike space and time, consciousness does not extend but embeds within timespace as an energetic substrate. While prior work has considered altered states (dreaming, psychedelics) as gateways to emergent particles such as the Trion, this paper extends the framework to planetary-scale coherence: the capture and redistribution of collective thought-energy.

Mindforce Capture is proposed as a mechanism whereby the dissipation of human thought into cosmic space can instead be recycled into a global coherent field. Such a process would equalize the distribution of informational energy (arts, culture, knowledge, medicine, science) and produce macroscopic physical consequences, including light modulation and shielding from extraterrestrial hazards.

2. Conceptual Foundation: The Fifth Dimension as Consciousness

Within BUFT, consciousness is treated not as a byproduct of neural activity but as a fundamental dimension of reality. This fifth dimension, called Mindforce, does not behave spatially and cannot be mapped onto conventional coordinates of length or volume. Instead, it is an energetic substrate embedded within the fabric of time and space, one that provides coherence and meaning to the flows of energy and matter.

Mindforce manifests through emanations of thought. Each human thought generates a ripple in this fifth dimension, described as quanta that couple with time (Thoton) and electromagnetism (Photon). In everyday experience, these emanations disperse freely into the universe, slowly dissipating and contributing little to the maintenance of collective order. The postulate of **Mindforce Capture** proposes a reversal of this process.

Rather than allowing thought-energy to radiate outward unchecked, collective consciousness can be aligned into a coherent resonance that traps and redirects these emanations.

The mechanism of capture is analogous to forming a planetary-scale standing wave. When consciousness enters a phase-aligned state across populations, the Mindforce field becomes self-reinforcing, generating a coherent structure that can persist over time. This coherence transforms dissipative mental energy into a stable planetary resource. The result is a new kind of field surrounding Earth—a fifth-dimensional analogue of the magnetic field—capable of redistributing informational energy and interacting with electromagnetic and gravitational structures. In this sense, the Mindforce dimension provides not only a theoretical foundation for consciousness but also a physical mechanism for global energy balancing and shielding.

3. Mathematical Framework

Governing Equation for Mindforce Capture

$$\mathscr{M}_{\mu\nu}\hbar^{\mu}\hbar^{\nu} = 0\,, \quad \textit{with} \quad \mathscr{M}_{\mu\nu} = g_{\mu\nu} + \Lambda(\Psi)\, u_{\mu}u_{\nu}$$

where:

$\mathscr{M}_{\mu\nu}$ – Mindforce metric (effective geometry of consciousness-modulated optics).

$g_{\mu\nu}$ – background BUFT timespace metric.

$\Lambda(\Psi)$ – Dimensionless coherence of the consciousness phase field Ψ.

u_{μ} – observer 4-velocity defining the frame of measurement.

$\hbar^{\mu}$ – wave-vector of observed light or Mindforce quanta (null trajectory).

Interpretation

Light and thought-quanta are null with respect to the Mindforce-modified geometry) $\mathscr{M}_{\mu\nu}$. Collective coherence modifies global optical conditions, redistributing both electromagnetic radiation and conscious energy.

4. Mathematical Validation

4.1. Null preservation:

Because $\nabla(\mathscr{M})$ is compatible with $\mathscr{M}_{\mu\nu}$, the null condition $\mathscr{M}_{\mu\nu}\,k^{\mu}\,k^{\nu} = 0$ is preserved along geodesics. This ensures that Mindforce quanta remain observable as "lightlike" entities in the fifth dimension.

4.2. Dimensional consistency:

All terms in $\mathscr{M}_{\mu\nu}$ are dimensionless modifications of the base metric, guaranteeing structural consistency in BUFT units.

4.3. Classical limit:

If $\Lambda(\Psi) = 0$ (no global coherence), then $\mathscr{M}_{\mu\nu} = g_{\mu\nu}$ and the system reduces to ordinary BUFT dynamics.

4.4. Predictive consequence:

Where $\Lambda(\Psi) \neq 0$, optical trajectories and Mindforce fluxes bend coherently, yielding observable redistribution in light scattering and aurora-like global luminescence.

5. Empirical Analogues

Although Mindforce Capture operates in the fifth dimension, its consequences may be illuminated by several terrestrial analogues. The aurora borealis provides one such analogy: solar particles guided by Earth's magnetic field into the polar regions release luminous energy that spreads across the sky. In the same way, Mindforce Capture channels the dissipative quanta of thought into coherent planetary streams, producing a distributed luminous effect—a fifth-dimensional aurora not limited to the poles but encompassing the entire planet.

Another parallel can be drawn with the Schumann resonances, the global electromagnetic standing waves trapped between Earth's surface and ionosphere. These resonances demonstrate the planet's natural ability to sustain coherent, planet-wide oscillations. If electromagnetic waves can self-organize into such patterns, then consciousness quanta, once captured, could similarly establish standing waves within the Mindforce field. These oscillations would serve as carriers for redistributed energy and information.

Finally, tentative empirical support arises from experiments in global consciousness research. Projects monitoring random number generators worldwide have reported statistical anomalies coinciding with periods of synchronized human attention, such as global events or crises. Though controversial, such findings hint at a latent capacity for collective consciousness to influence physical systems. BUFT provides a theoretical scaffolding for these results: rather than being statistical flukes, they may be the subtle precursors of large-scale Mindforce Capture.

Together, these analogues suggest that Mindforce Capture is not an isolated abstraction. Instead, it resonates with known natural processes that demonstrate coherence, redistribution, and field-like behavior at planetary scales.

6. Implications

6.1. Redistribution of Energy and Knowledge

If Mindforce Capture were realized, the flow of arts, culture, science, and medicine would no longer follow fragmented or uneven pathways. Instead, these informational energies would circulate through a coherent global lattice, one that recycles rather than dissipates thought-energy. This redistribution would operate as a planetary equalizer: knowledge and creativity generated in one region would be rapidly accessible in others, while barriers of isolation and entropic decay would diminish. Wealth and health, traditionally tethered to unequal distributions of information and resources, would begin to stabilize as emergent outcomes of this unified consciousness field. In essence, Mindforce Capture offers a mechanism by which humanity's shared intellectual and cultural achievements could continuously reinforce planetary well-being.

6.2. Effects on Light Distribution

Solar light entering Earth's system would refract differently under $\mathcal{M}_{\mu\nu}$

Result: softened global illumination analogous to a distributed aurora, without destructive intensity. Health risks minimal, since coherent redistribution reduces local peaks in EM stress.

6.3. Constraints and Challenges

The realization of Mindforce Capture depends critically on the possibility of achieving global coherence in human consciousness. Phase alignment across billions of individuals is a profoundly complex socio-physical problem, requiring not only scientific mechanisms of synchronization but also cultural, psychological, and ethical cooperation. Even if partial coherence were achieved, the stability of the resulting field would be vulnerable to decoherence, as local disruptions could destabilize planetary-scale resonance. Quantitative field-strength predictions depend on the calibrated coherence function $\Lambda(\Psi)$. Empirical characterization of this function supplies the basis for evaluating planetary redistribution and shielding.

These requirements define the program for quantitative testing of the proposed planetary field.

7. Visualization: A Thought Experiment

To imagine the perceptual consequences of Mindforce Capture, consider the sky as seen by a human observer once planetary coherence has been established. The firmament would no longer present itself as a uniform blue dome by day or a star-punctured darkness by night. Instead, subtle waves of iridescence would ripple across the atmosphere, resembling the aurora yet far more evenly spread, extending from horizon to horizon. These patterns would not be dazzling or overwhelming but rather softly luminous, their shifting hues synchronized with the flux of collective thought-energy.

At times of heightened global coherence, the sky would appear to shimmer with faintly braided currents of light, as though the atmosphere itself were suffused with consciousness. Unlike conventional auroras localized to high latitudes, this effect would be planet-wide, perceptible to all observers regardless of geography. It would be gentle to the human eye—not an intrusive glare, but a diffused and living texture in the heavens. Such a view would serve as both aesthetic confirmation of Mindforce Capture and empirical evidence of the fifth-dimensional coherence underpinning it.

8. Pathways to Realization and Testable Prediction

The practical achievement of Mindforce Capture using today's technology would require coordinated coupling between human consciousness and electromagnetic infrastructures. One pathway involves the integration of global meditation or synchronized attention practices with sensor networks already deployed for geophysical monitoring. Arrays of magnetometers, ionospheric probes, and random number generators could provide real-time feedback on coherence levels in collective consciousness, effectively acting as resonance detectors for Mindforce fluctuations.

Another feasible method is the use of satellite constellations designed to detect anomalous phase correlations between electromagnetic signals and periods of synchronized global events. By combining human participation (mass attention or meditation at designated times) with continuous atmospheric and space-based monitoring, it becomes possible to test whether Mindforce coherence produces measurable deviations in light scattering, plasma distribution, or Schumann resonance amplitudes.

Testable Prediction: If Mindforce Capture is valid, then during globally synchronized consciousness events (e.g., coordinated meditation periods involving millions of participants), the Schumann resonance amplitudes should exhibit statistically significant, non-random coherence spikes that cannot be explained by solar or geomagnetic activity. These spikes would occur simultaneously across geographically distant monitoring stations, providing evidence of a fifth-dimensional organizing principle coupling thought-energy with electromagnetic oscillations.

—※—※—※—

References

1. Bakous, R. (2025). Bakous Unified Field Theory: A Timespace Perspective. Independently Published.

2. Barceló, C., Liberati, S., and Visser, M. (2011). Analogue Gravity. Living Reviews in Relativity, 14, Article 3. https://doi.org/10.12942/lrr-2011-3

3. Schumann, W. O. (1952). Über die strahlungslosen Eigenschwingungen einer leitenden Kugel, die von einer Luftschicht und einer Ionosphärenhülle umgeben ist. Zeitschrift für Naturforschung A, 7(2), 149–154. https://doi.org/10.1515/zna-1952-0202

4. NASA. (n.d.). Auroras. NASA Science. https://science.nasa.gov/sun/auroras/ (accessed 26 October 2026).

5. Nelson, R. D., Radin, D. I., Shoup, R., and Bancel, P. A. (2002). Correlations of continuous random data with major world events. Foundations of Physics Letters, 15, 537–550. https://doi.org/10.1023/A:1023981519179

6. Bancel, P. A. (2017). Searching for Global Consciousness: A 17-Year Exploration. Explore, 13(2), 94–101. https://doi.org/10.1016/j.explore.2016.12.003

These references document the BUFT foundation, effective-metric methods, auroral physics, Schumann resonances, and collective-attention research discussed in Article 41.

ARTICLE 42

Perceptual Time Dilation Through Neural Compression and Conscious Resonance: A BUFT-Based Postulate

ABSTRACT

This paper introduces a novel postulate within the Bakous Unified Field Theory (BUFT) asserting that human observers can artificially induce subjective time dilation while biologically awake by modulating internal harmonic coherence and reducing entropy-driven cognitive interference. Time, under BUFT, manifests as a kinetic field through the continuous harmonic propagation of energy into mass at the speed of light via $E = mc^2$ *in the Bakous Energy Field (BEF).*

While timespace curvature governed by gravity determines the universal structure of temporal flow, the observer's ability to experience time is further modulated by local entropy conditions within the conscious field. By reducing neural information distance and enhancing internal resonance, it becomes possible to perceptually expand time without altering its objective passage.

Neural implants, pharmacological agents, and entrainment devices can potentially stabilize this internal resonance, allowing observers to access temporally expanded cognitive states within ordinary waking life, effectively enriching perception, learning, and awareness within a finite biological lifespan.

1. Introduction

The human experience of time is variable. In dreams, a single second may encompass a narrative of hours; in states of boredom, minutes feel interminable; and in moments of joy, hours vanish in an instant. These are not psychological tricks—they are signatures of a deeper phenomenon. Within BUFT, gravity governs forward temporal progression. Consciousness and local entropy modulate the observer's temporal experience within that continuous progression.

This paper proposes a bold extrapolation of this principle: that time, as experienced by the human observer, can be artificially modulated by altering internal entropy structures and field harmonic coherence. By compressing the information distance across neural systems and stabilizing timespace curvature within the conscious field, it is possible to experience an expansion of subjective time—effectively increasing mental presence, learning capacity, and perceptual resolution within the same objective duration.

2. Theoretical Framework: Time and Neural Propagation in BUFT

2.1. Time as a Kinetic Field and the Role of Thoton and Choton

In BUFT, time is a kinetic field—a dynamic expression of energy propagating through the Bakous Energy Field (BEF) as a result of energy-mass equivalence described by $E = mc^2$. This equation, foundational to the BUFT cosmology, is not merely a physical identity but a generative function: energy condenses into mass, generating observable timespace structure as harmonic field derivatives. From this interaction emerge the two fundamental quantum carriers of time-conscious dynamics: the Thoton and the Choton.

- **The Thoton** is the quantum particle responsible for propagating the flow of time through curved timespace. It acts as the kinetic agent of temporal motion, ensuring the continuity and directionality of experience across the BEF.
- **The Choton** mediates the interaction between consciousness and the temporal field. It is through Choton entanglement that observation anchors time into perceivable intervals. **The Choton** does not create time, but modulates the observer's field coupling to the kinetic flow generated by the Thoton.

This dyadic structure—Thoton *as the vector, Choton as the tuner—defines* how consciousness rides the river of time. Time is not made by observation but is observed as a consequence of BEF dynamics. However, the manner in which time is sampled, resolved, and experientially sequenced is subject to the configuration of the conscious field.

2.2. Temporal Expansion via Neural Compression

The human brain, within BUFT, is not simply a biological processor but a localized resonator of the BEF. The structure and topology of this resonator—how far signals must travel across the cortex and how much entropy must be processed to maintain awareness—determine the resolution and velocity of experienced time.

When neural information distance is reduced—by enhancing signal coherence or minimizing routing complexity—the effective propagation delay between cognitive regions is shortened. As a result, more complete perceptual and analytical loops can occur per unit of subjective time. This does not make cognition faster in the conventional sense. Instead, it creates the conditions for time to dilate internally, making each moment denser with experience and intelligence.

Temporal awareness under neural compression becomes granular. As the temporal curvature within the cognitive field flattens, the observer gains access to increased perceptual threading—subroutines of thought that run in parallel and in depth. Learning accelerates, not by external urgency, but by internal spaciousness. Awareness becomes high-resolution. The mind shifts from processing events in linear fragments to absorbing them in layered totalities. Time does not stretch—it deepens.

3. Artificial Induction of Subjective Time Dilation

3.1. Methods of Temporal Modulation

This section outlines engineered approaches to induce subjective time dilation in the conscious human observer. Each is grounded in BUFT and designed to replicate naturally occurring low-entropy, high-coherence states through artificial means.

1. Neural Entrainment via External Devices

Advanced neural helmets or implants can be developed to emit structured, low-entropy harmonic signals that align with the brain's internal field resonance. By carefully controlling frequency, phase coherence, and amplitude modulation, these devices stabilize the local time-field curvature without disrupting consciousness.

Such entrainment devices do not induce sleep or trance. Instead, they replicate the oscillatory precision of deep sleep while maintaining alertness, allowing the user to remain fully awake while experiencing dilated time. This is achieved through sympathetic resonance, wherein the brain locks into external harmonics, reducing internal entropic flux.

Testable Prediction: Subjects under entrainment should report extended subjective duration during standardized cognitive tasks, accompanied by increased gamma/theta coherence and lower entropy signatures in neural scans.

2. Pharmacological Temporal Modulators

Under this postulate, pharmacological modulation targets serotonergic and GABAergic pathways to stabilize the Choton–Thoton interface. The proposed modulators reduce neural signaling noise and support harmonic coherence while sustaining lucid awareness.

Under these conditions, the brain can achieve a state similar to cognitive stasis—informationally calm yet perceptually lucid. Time dilates not because of psychotropic disruption, but because entropy filtration is **enhanced, and neural information distance is functionally minimized.**

Testable Prediction: Controlled administration of entropy-reducing compounds should yield measurable reductions in reaction time variability, increases in perceived task duration, and elevated signal-to-noise ratios in cortical data processing.

3.2. Resonance Locking and Temporal Phase Stability

Artificial time dilation depends not solely on compressing information distance or damping entropy but on achieving resonance locking—a stabilized energetic coupling between the observer's conscious field and the harmonics of the Bakous Energy Field (BEF). When this coupling reaches equilibrium, the temporal phase of perception enters a coherently extended regime. This is temporal phase stability, wherein subjective time flows more slowly and richly, yet consciousness remains dynamically engaged and alert.

Resonance locking is accomplished when internal neural harmonics and entropy damping mechanisms sustain a phase-aligned relationship with the Thoton-driven temporal field. This results in uninterrupted temporal dilation that can persist as long as energetic balance is maintained.

Testable Prediction: In experimental settings where subjects are subjected to resonance-inducing harmonic stimuli (via neural entrainment devices), EEG spectral analysis should reveal sustained, narrow-band coherence at theta-gamma cross-frequencies.

Simultaneously, subjects should report a reliable and repeatable elongation of time perception during controlled-duration tasks when compared to baseline sessions without resonance locking. The effect should scale with the degree of phase stability observed in the neural signatures.

4. Mathematical Model

The core subjective dilation function within the BUFT framework is:

$$\hat{T}(t) = \tau\, e^{\frac{-\gamma S(t)\, d_{\mathrm{n}}(t)}{d_{\mathrm{n},0}^{2}}}$$

where:

$\hat{T}(t)$ — Subjective temporal scale characterizing experienced duration at time t.

τ — Baseline temporal resolution of the conscious system.

γ — Positive coupling making the perceptual exponent dimensionless.

$S(t)$ — Local entropy influx rate contributing to cognitive interference.

$d_{\mathrm{n}}(t)$ — Effective neural information distance set by signal routing across cognitive regions.

$d_{\mathrm{n},0}$ — Reference neural distance defining the baseline configuration.

t — Objective time indexing the evolving cognitive state.

Reduced entropy influx and shorter neural information distance increase the subjective temporal scale while objective temporal progression remains continuous.

5. Implications for Cognition, Learning, and Lifespan

5.1. Cognitive Density and Conscious Throughput

At the core of BUFT's implications is the principle of **cognitive density—the** capacity to saturate each perceptual moment with layers of meaning, analysis, and recall. As neural compression minimizes signal delay and entropy reduction quiets irrelevant stimuli, the bandwidth of working consciousness expands. This is not simply multitasking, but multi-scalar thinking: the ability to operate simultaneously across macro- and micro-cognitive levels.

A single idea can be examined from multiple angles at once. A memory can be relived not as data, but as an internally reconstructed, time-rich experience. Strategic reasoning and creative flow coalesce, no longer separated by mental fatigue or linear limitations. In this state, **conscious throughput—the** volume and complexity of thoughts per unit of subjective time—reaches unprecedented saturation.

This also redefines the concept of expertise. Mastery becomes not merely the accumulation of information, but the capacity to inhabit information temporally, allowing the thinker to stretch, analyze, and manipulate thoughtforms across a personally dilated mental canvas.

5.2. Conscious Temporal Sculpting and the Architecture of Memory

Temporal sculpting is the act of consciously modulating time perception to match cognitive demands. In the BUFT model, this is no longer abstract. Neural implants, pharmacological support, and field-based harmonics provide tools for real-time dilation adjustment—

allowing the thinker to slow internal time during complex tasks or accelerate it when appropriate.

This ability revolutionizes memory. Learning becomes layered into **chronoform structures—time-textured** experiences rather than snapshots. The emotional and logical architecture of memories becomes retrievable not only by content but by temporal rhythm. This endows recall with cinematic fidelity: memories are no longer just reactivated—they are reinhabited.

Furthermore, this architecture allows memory to self-organize not by category or priority, but by chronotonal resonance: the intrinsic frequency signature of the memory field. This reorganizes knowledge networks to optimize conceptual fusion, creativity, and innovation.

5.3. Experiential Immortality and the Expansion of Subjective Lifespan

Experiential immortality under BUFT is defined not by the biological extension of life but by the expansion of lived time within a fixed lifespan. This temporal amplification is made possible by sustained harmonic resonance within the conscious field, allowing an individual to experience the equivalent of lifetimes within the frame of a single biological existence.

In practice, a person living to 80 years of age could, through daily engagement in resonance-locked states, subjectively experience the richness and depth of several centuries of reflective, creative, or intellectual life. Each day becomes a container for weeks of insight. Each moment unfolds into a gallery of attention, emotion, and processing.

Examples of perceived immortality through subjective dilation include:

- **The Terminal Mind Expansion**: A patient facing the end of life could spend what externally amounts to hours in a temporal chamber but internally live through months of closure, expression, and peace. Grief transforms into narrative. Farewells become symphonies.

- **The Scholar's Epoch:** A researcher could, in deep harmonic states, complete decades' worth of intellectual synthesis, analysis, and innovation in mere calendar months. An entire school of thought could be built in the architecture of internal time.
- **The Artistic Bloom:** An artist working in a field-compressed resonance state could create a vast body of internally experienced work during a brief residency, distilling lifetimes of emotional and aesthetic development into a single masterpiece.

Thought Experiment—The Chrono-Vault

Imagine entering a controlled chamber—a chrono-vault—designed to achieve maximum resonance locking through an optimized combination of pharmacological tuning, harmonic field entrainment, and neural phase compression. You sit for one hour.

Inside this hour, you live an entire subjective year. You read, reflect, feel, create, and discover. You do not age externally, but internally you emerge changed—as if you had lived through 365 days of conscious engagement.

Upon returning to the outside world, only an hour has passed. To others, you were briefly still. To yourself, you have lived richly and deeply. Your thoughts are matured, your insights complete, your relationships freshly understood.

This is not science fiction. Under BUFT, it is the logical consequence of manipulating entropy, neural distance, and harmonic alignment within the architecture of time.

The Result: Time becomes art. Memory becomes sculpture. Life becomes deeper, not longer. And immortality, in its purest form, is reclaimed not through resisting death—but through expanding the very substance of living.

6. Conclusion

Under the Bakous Unified Field Theory, time manifests as a kinetic field through the harmonic propagation of energy into mass via $E = mc^2$ in the Bakous Energy Field. Gravity governs continuous forward temporal progression, while consciousness modulates its experience through local entropy conditions and neural compression. Subjective time can therefore be sculpted while objective progression remains continuous. By managing entropy flow and compressing neural distance, human consciousness can experience time in expanded resolution. This preserves time's continuity while enriching its occupancy. Through harmonic resonance and temporal sculpting, the observer is empowered not merely to measure time, but to live more deeply within it.

References

1. Bakous, R. N. (2025). *Bakous Unified Field Theory: A Time-Space Perspective.*
2. Carhart-Harris, R. L., & Friston, K. J. (2019). *REBUS and the Anarchic Brain: Toward a Unified Model of the Brain Action of Psychedelics.* Pharmacological Reviews, 71(3), 316–344.
3. Tononi, G., & Edelman, G. M. (1998). *Consciousness and Complexity.* Science, 282(5395), 1846–1851.
4. Hameroff, S. R., & Penrose, R. (2014). *Consciousness in the Universe: A Review of the 'Orch OR' Theory.* Physics of Life Reviews, 11(1), 39–78.

—※—※—※—

ARTICLE 43

Entropy as the Sixth Dimension in the Bakous Unified Field Theory: Exploring the Role of Entropy in the Temporal-Spatial Continuum

ABSTRACT

The Bakous Unified Field Theory (BUFT) integrates entropy as a sixth dimension. Within the BUFT framework, entropy is posited as a dynamic dimension that emerges through the interplay of gravity, time, and electromagnetism. By incorporating entropy as an intrinsic property of timespace, BUFT connects energy dispersion with the forward temporal progression governed by gravity.

This paper develops a mathematical representation of entropy as a dimension and examines its relationship to the cosmic microwave background, black hole thermodynamics, and conscious observation. These phenomena provide physical context for the BUFT postulate and its dimensional representation.

Introduction

The foundation of BUFT is the postulate that the universe originated from a quantum fluctuation within a one-dimensional state of pure energy and pure matter, expanding at the speed of light. This fluctuation, governed by the mass–energy relation $E = mc^2$, propagated timespace and matter in a harmonic interplay of oscillating strings. These strings generated space through successive initiations of energy–matter transformations, perpetuating larger spatial expansions.

This framework inherently links entropy to the forward movement of time, as governed by gravity. In this context, entropy is not merely a measure of disorder but a dimension that records irreversible changes within timespace. Gravity governs temporal progression; entropy follows that progression. This paper develops the representation of entropy as the sixth dimension within BUFT and examines its relationship to observable phenomena.

MATHEMATICAL FRAMEWORK

1. Entropy and the Arrow of Time

Entropy (S) measures energy dispersion relative to timespace. For equally probable accessible microstates, the Boltzmann expression is:

$$S = k_B \ln \Omega(t)$$

S — Entropy associated with the accessible microstates of the system.

k_B — Boltzmann constant relating microstate multiplicity to entropy.

$\Omega(t)$ — Number of accessible microstates at time t, treated as a positive differentiable function in the continuum description.

Differentiating the entropy expression with respect to time gives:

$$\frac{dS}{dt} = \frac{k_B}{\Omega(t)} \frac{d\Omega(t)}{dt}$$

dS/dt – Rate of change of the system's entropy along its temporal evolution.

dΩ(t)/dt – Rate of change of the accessible microstate count in the continuum description.

t – Temporal progression coordinate.

This identity relates entropy change to changing microstate accessibility. Within BUFT, gravity governs forward temporal progression. For energy-redistribution entropy, the inverse relationship with gravitational intensity applies at fixed encoding and temporal-rate factors; the inverse scaling is specified by the BUFT coupling relation.

2. Entropy as a Dimension

In BUFT, entropy is incorporated into the timespace metric as the sixth dimension (x_6), augmenting the conventional four dimensions (x_1, x_2, x_3, t) and the fifth dimension of consciousness (x_5). With the additional coordinates expressed on length scales, the local metric form is:

$$ds^2 = -c^2dt^2 + dx_1^2 + dx_2^2 + dx_3^2 + dx_5^2 + dx_6^2$$

Here, x_6 is proportional to accumulated entropy change, with a conversion factor that preserves the metric's length scale:

$$x_6(t) = x_{6,0} + \alpha_6 \int_{t_0}^{t} \frac{dS(u)}{du}\, du$$

$$x_6(t) = x_{6,0} + \alpha_6[S(t) - S(t_0)]$$

$x_6(t)$ — Sixth coordinate representing accumulated entropy change.

$x_{6,0}$ — Reference value of the sixth coordinate at t_0.

α_6 — Constant conversion factor from entropy change to sixth-coordinate displacement.

t_0 — Reference time defining the initial entropy and coordinate values.

u — Integration variable spanning the temporal interval from t_0 to t.

c — Speed of light in vacuum.

This definition fixes the entropy-to-coordinate conversion. BUFT postulates light-speed propagation; the coordinate definition specifies the associated geometry.

3. Entropy and Conscious Observation

Conscious observation, as described in BUFT, is an electromagnetic phenomenon permanently encoded into timespace. BUFT links this permanent encoding to energy conservation, retaining thought-energy within the universe. The associated entropic encoding is distinct from energy-redistribution entropy.

EMPIRICAL CONTEXT

1. Cosmic Microwave Background (CMB)

The nearly isotropic CMB records the early universe's thermal state. Its distribution provides observational context for studying cosmic thermal evolution. Isotropy characterizes the observed distribution; BUFT interprets entropic encoding through the sixth coordinate.

2. Black Hole Thermodynamics

Black hole thermodynamics relates entropy to the surface area of the event horizon:

$$S_{BH} = \frac{k_B c^3 A}{4G\hbar}$$

S_{BH} – Bekenstein–Hawking entropy of the black hole.

A – Surface area of the event horizon.

G – Newton's gravitational constant.

$\hbar$ – Reduced Planck constant.

This relationship connects entropy with horizon geometry. Within BUFT, it is interpreted through the temporal-boundary account of the horizon. The area relation provides physical context for BUFT's sixth-dimension interpretation.

CONSTRAINTS AND OVERCOMING THEM

Constraint 1: Observational Limitations

Human observation is inherently constrained by the inability to directly perceive higher dimensions. BUFT addresses this by relating entropy's effects to observable phenomena, such as the CMB and black hole thermodynamics.

Constraint 2: Mathematical Representation

Incorporating entropy into the timespace metric requires an explicit conversion between entropy change and coordinate displacement. The conversion factor α_6 supplies this scale; the relation to gravitational and electromagnetic dynamics remains part of the BUFT model.

Conclusion

Entropy, as a dimension within the Bakous Unified Field Theory, provides a framework for relating energy dispersion, conscious encoding, and the evolution of the universe. Its metric representation preserves gravity as the governor of forward temporal progression and entropy as following that progression.

The sixth-dimension postulate connects these processes within BUFT. The CMB and black hole thermodynamics provide physical context, while experimental tests must distinguish this dimensional interpretation from existing descriptions.

Future Work

High-precision measurements of entropy gradients in gravitational fields and simulations of timespace curvature can support tests of the proposed relationship between entropy and the sixth coordinate.

References

1. Bardeen, J. M., Press, W. H., and Teukolsky, S. A. (1972). Rotating black holes: Locally nonrotating frames, energy extraction, and scalar synchrotron radiation. Astrophysical Journal, 178, 347–369.

2. Remillard, R. A., and McClintock, J. E. (2006). X-ray properties of black-hole binaries. Annual Review of Astronomy and Astrophysics, 44, 49–92.

3. Abbott, B. P., et al. (LIGO Scientific Collaboration and Virgo Collaboration). (2016). Observation of gravitational waves from a binary black hole merger. Physical Review Letters, 116(6), 061102.

4. Penrose, R. (1969). Gravitational collapse: The role of general relativity. Rivista del Nuovo Cimento, 1, 252–276.

5. Gammie, C. F., McKinney, J. C., and Tóth, G. (2003). HARM: A numerical scheme for general relativistic magnetohydrodynamics. Astrophysical Journal, 589, 444–457.

—※—※—※—

Article 44

Entropy as the Sixth Dimension: The Conservation of Conscious Thought in Time-Space

Abstract

The Bakous Unified Field Theory (BUFT) redefines the foundational nature of reality by positing timespace as the fundamental medium through which all interactions manifest, governed by the Bakous Energy Field (BEF). Within this framework, entropy is elevated beyond a mere statistical measure of disorder to a spatial dimension—the sixth dimension—responsible for encoding fluctuations in conscious observation and electromagnetic interactions. Conscious thought, as the observation of electromagnetism, must be conserved within timespace to uphold the fundamental law of energy conservation. This paper introduces a mathematical model in which the entropy dimension provides a structured repository for these cognitive interactions, ensuring that no information is lost but rather redistributed within time space. The model builds upon principles from quantum mechanics, thermodynamics, and gravitational physics to establish a formal description of how conscious energy is encoded within the fabric of reality. Experimental predictions, including the impact of entropic encoding on gravitational wave signatures, quantum coherence, and black hole entropy, provide potential avenues for future validation.

1. Introduction

1.1. The Need for an Extended Dimensional Framework in BUFT

BUFT presents a vision of timespace as a dynamic, self-regulating structure in which all interactions emerge as manifestations of its underlying geometry. Unlike conventional relativistic models that treat space and time as a unified four-dimensional continuum, BUFT asserts that timespace, governed by the Bakous Energy Field (BEF), is the fundamental medium shaping the evolution of matter, energy, and consciousness. Within this framework, the passage of time is not an emergent consequence of spatial interactions but a primary force that dictates the structure of reality itself.

Entropy, long regarded as a statistical measure of disorder, is reinterpreted as an intrinsic spatial coordinate—an extension of timespace that encodes transformations of energy and information. The fluctuations of entropy are not merely passive reflections of physical processes but active geometric distortions within the topology of timespace. If energy conservation is to remain fundamental, then conscious observation, as an interaction with the electromagnetic field, must also be preserved within this structure. The sixth dimension provides a means by which thought energy, once dissipated, is encoded and stored without violating the foundational laws governing timespace dynamics.

2. Entropic Encoding and the Conservation of Thought-Energy

The Bakous Energy Field (BEF) serves as the fundamental medium through which all interactions in timespace occur, dictating the evolution of energy, matter, and consciousness. In this framework, conscious observation—defined as the act of collapsing electromagnetic potential states—is not a passive event but an active transformation that must be accounted for within the conservation laws of time and space. To ensure that the energy associated with conscious thought is not lost, BUFT posits that it is transcribed into the entropy dimension, where it is structured and preserved.

Rather than fading into randomness, thought energy undergoes an ordered transition into the entropy coordinate, ensuring that no information is arbitrarily dissipated. Unlike classical thermodynamic entropy, which merely quantifies disorder, entropic encoding in BUFT operates as a structured transformation in which cognitive interactions leave geometric imprints within time space. This process enables the conservation of conscious energy by redistributing it within a spatial framework rather than allowing it to vanish into nonexistence.

This mechanism suggests that entropy, far from being an abstract statistical function, is an active dimension that shapes the evolution of reality. Each act of observation subtly alters the fabric of timespace, embedding its influence into the entropic coordinate. The structured nature of this encoding implies a form of information persistence, where conscious interactions contribute to the evolving topology of reality itself. The function of entropy in BUFT is, therefore, not merely to measure transformation but to facilitate a deeper continuity of information, integrating thought energy into the very architecture of the universe.

3. Mathematical Formulation of the Encoding Process

3.1. Time-Space Metric with Entropy as a Dimension

In BUFT, the fundamental metric governing timespace is extended to incorporate the entropy dimension:

$$ds^2 = -c^2 dt^2 + dx_1^2 + dx_2^2 + dx_3^2 + dx_5^2 + f(x_6)dx_6^2$$

where:

ds^2 — Squared infinitesimal interval in six-dimensional timespace.

c — is the speed of light,

dt — denotes the differential time coordinate,

dx_1, dx_2, and dx_3 — correspond to the conventional spatial dimensions,

dx_5 — represents the consciousness dimension,

dx_6 — is the entropy coordinate,

$f(x_6)$ — is a function that modulates the influence of entropic distortions on timespace curvature.

3.2. Modified Einstein Equations with the BEF Influence

The Einstein field equations, adapted to accommodate entropic encoding, take the form:

$$R_{\mu\nu} - \frac{1}{2} g_{\mu\nu} R + \Lambda g_{\mu\nu} = \frac{8\pi G}{c^4} \left(T_{\mu\nu} + \Theta_{\mu\nu}\right)$$

where:

$R_{\mu\nu}$ — represents the Ricci curvature tensor,

$g_{\mu\nu}$ — is the metric tensor describing timespace geometry,

R — is the Ricci scalar,

Λ — is the cosmological constant,

G — is Newton's gravitational constant,

$T_{\mu\nu}$ — is the energy-momentum tensor,

$\Theta_{\mu\nu}$ — Entropic energy–momentum contribution in the field equation.

4. Experimental and Observational Considerations

4.1. Entropic Imprints on Gravitational Wave Signatures

The structured encoding of cognitive energy within the entropy dimension implies that gravitational waves propagating through high-entropy regions of timespace may exhibit measurable distortions. If the redistribution of thought energy follows the formalism described in Section 3, gravitational wave interferometers may detect anomalous shifts in waveform attenuation and frequency dispersion. The predicted effect would manifest as a subtle but consistent deviation from classical general relativity predictions, detectable through high-precision instruments such as LIGO and Virgo.

4.2. Quantum Coherence and the Persistence of Thought Information

If entropic encoding operates as theorized, then quantum systems influenced by observation should exhibit prolonged coherence lifetimes. Entangled quantum states subjected to direct human observation may experience modified decoherence rates due to their interaction with the entropy dimension. Experimental tests involving delayed-choice quantum erasure setups could provide empirical data to support or refute this hypothesis, offering insight into the relationship between consciousness and quantum mechanics.

4.3. Black Hole Information Storage and the Sixth Dimension

If entropy exists as a fundamental spatial coordinate, then black holes must serve as gateways for information displacement into the entropy dimension. Instead of information being lost upon crossing the event horizon, it is restructured within the sixth-dimensional topology, ensuring that conservation laws remain intact. Hawking radiation, in this model, may serve as an entropic retrieval mechanism, gradually releasing stored information back into observable timespace. This hypothesis aligns with the holographic principle and offers a resolution to the black hole information paradox, suggesting that black-holes act not as absolute erasers of information, but as conduits for entropic redistribution.

5. Conclusion

This paper extends the Bakous Unified Field Theory (BUFT) by introducing entropy as a spatial dimension responsible for encoding conscious thought while preserving the conservation of energy. The formalism presented describes the process by which cognitive energy is transferred into the entropy dimension, ensuring that no information is lost but is redistributed within time space. Experimental predictions, including the influence of this entropic encoding on gravitational wave signatures, quantum coherence, and black hole entropy, provide avenues for future research and potential experimental verification. By grounding consciousness in the topology of timespace, this framework offers new insights into the fundamental structure of reality.

References

1. Bekenstein, J. D. (1973). *Black holes and entropy.* Physical Review D, 7(8), 2333-2346.
2. Hawking, S. W. (1975). *Particle creation by black holes.* Communications in Mathematical Physics, 43(3), 199-220.
3. Penrose, R. (1994). *Shadows of the Mind: A Search for the Missing Science of Consciousness.* Oxford University Press.
4. Susskind, L. (1995). *The world as a hologram.* Journal of Mathematical Physics, 36(11), 6377-6396.

—※—※—※—

Article 45

Validation of Time Force as a Distinct Phenomenon from Entropy in the Bakous Unified Field Theory Framework

Abstract

The Bakous Unified Field Theory (BUFT) describes time-force as possessing kinetic energy and wave-like properties, distinct from entropy, which BUFT identifies as the sixth dimension. Within this framework, time-force propagates via massless particles termed Thotons, while energy-redistribution entropy follows gravitationally governed temporal progression. Gravitational radiation is a propagating curvature disturbance and is distinct from entropy.

This paper formulates the distinction between time-force and entropy and develops experimental proposals for testing it. The inverse relationship between time-force and magnetic field strength, the role of gravity in forward temporal progression, and the interaction between electromagnetism and conscious observation organize the analysis. Proposed quantum, neurophysiological, and cavity measurements provide routes for investigating the BUFT account.

1. Introduction

Conventional physics treats time as a coordinate or parameter; it does not generally derive time's existence from entropy. BUFT postulates that time is an active force—time-force—with wave-like properties and kinetic energy. Unlike energy-redistribution entropy, which characterizes energy dispersion within gravitational fields, time-force is postulated to propagate via Thotons, massless particles associated with its wave-like character.

This paper examines time-force as a distinct phenomenon from entropy. We explore its proposed kinetic and wave-like properties, its relationship with gravity and electromagnetism, and experimental frameworks for testing its fundamental nature.

2. Theoretical Foundations of Time Force and Entropy

2.1. Time Force: A Wave with Kinetic Energy

BUFT proposes a time-wave field satisfying the free-wave equation:

$$\nabla^2 T_f - \frac{1}{c^2}\frac{\partial^2 T_f}{\partial t^2} = 0$$

T_f — Time-wave field amplitude, distinct from the temporal coordinate.

∇^2 — Spatial Laplacian acting on the time-wave field.

t — Temporal progression coordinate.

c — Speed of light in vacuum.

This equation represents oscillatory propagation at c in the free-wave regime. BUFT postulates the kinetic interpretation; the field-energy relation and particle model specify its physical content.

Furthermore, BUFT relates time-force to magnetic field strength through an inverse proportionality:

$$F_i \propto \frac{1}{B}$$

F_i — Magnitude of time-force in the stated inverse magnetic relationship.

B — Magnitude of the magnetic field in the region considered.

At fixed remaining model factors, stronger magnetic fields correspond to reduced time-force in this relationship. It applies over the nonzero-field regime in which the proportionality is assumed; the expression does not prescribe the zero-field limit.

2.2. Entropy as a Non-Kinetic Sixth Dimension

Energy-redistribution entropy (S) in BUFT characterizes the redistribution of energy within a gravitational field. Unlike time-force, entropy is not itself a kinetic-energy quantity or a fundamental particle carrier. For the special case of two equally probable microstates, its value is:

$$S = k_B \ln(2)$$

S — Entropy of the system with two equally probable accessible microstates.

k_B — Boltzmann constant relating microstate multiplicity to entropy.

The factor ln(2) specifies the binary configuration. This fixed value is not a law of linear entropy growth; changing microstate multiplicity requires the general Boltzmann expression given in Article 43.

Entropy's role as the sixth dimension further distinguishes it from time-force. Energy-redistribution entropy follows temporal progression rather than driving it. Gravity, expressed as timespace curvature, governs forward temporal progression. Permanent entropic encoding of thought, discussed in Articles 43 and 44, is distinct from energy-redistribution entropy.

3. Empirical Tests and Thought Experiments

3.1. Experimental Distinction Between Time Force and Entropy

To test BUFT's distinction between time-force and entropy, we propose comparing time-dependent quantum phases under varying gravitational conditions:

1. Hypothesis: A time-wave contribution could produce phase oscillations beyond those predicted from ordinary quantum evolution and gravitational time dilation.

2. Setup: Apply a high-frequency electromagnetic field to an isolated quantum system while varying the gravitational conditions and monitoring temperature, environmental coupling, and phase stability.

3. Test criterion: Compare the measured phases with the conventional quantum and relativistic baseline and a specified BUFT contribution. Time-force is tested through phase-coupling predictions.

3.2. The Consciousness–Time Connection and Sleep-State Tests

BUFT postulates that conscious observation, tied to electromagnetism in the brain, interacts with time-force. Neurophysiological observations can investigate the proposed relationship:

Prediction: Changes in neural electromagnetic activity across sleep stages are associated with changes in subjective temporal experience. Deep sleep retains neural activity; objective temporal progression remains continuous.

Experimental setup: Use magnetoencephalography (MEG) to compare neural magnetic fluctuations across sleep stages with assessed temporal experience. MEG tests the proposed neural relationship, not time-wave propagation speed or cessation of objective progression.

4. Resolving Constraints in BUFT

The Role of Gravity in Time's Forward Motion

BUFT postulates that gravity, expressed as timespace curvature, governs light propagation and enables forward temporal progression. A proposed effective extension of the Einstein field equations includes a time-force energy–momentum contribution:

$$R_{\mu\nu} - \frac{1}{2}Rg_{\mu\nu} + \Lambda g_{\mu\nu} = \frac{8\pi G}{c^4}(T_{\mu\nu} + T^{(T)}_{\mu\nu})$$

$R_{\mu\nu}$ — Ricci curvature tensor of the effective four-dimensional metric.

$g_{\mu\nu}$ — Metric tensor describing the observable timespace geometry.

R — Ricci scalar.

Λ — Cosmological constant.

G — Newton's gravitational constant.

$T_{\mu\nu}$ — Energy–momentum tensor of matter and radiation.

$T^{(T)}_{\mu\nu}$ — Time-force energy–momentum tensor, with the same physical normalization as the matter and radiation source.

The indices μ and ν run over the observable four-dimensional coordinates. The combined source must satisfy covariant conservation. This proposed extension represents time-force within gravitational dynamics while retaining temporal progression as governed by curvature rather than entropy.

5. The Distinction Between Time Force and Gravitational Radiation

Gravitational radiation, energy-redistribution entropy, and time-force denote different quantities. Gravitational radiation consists of propagating disturbances of spacetime curvature.

Entropy characterizes microstate multiplicity and energy redistribution. BUFT additionally postulates a time-wave field carried by Thotons.

Entropy: For an isolated system, the second law constrains the total entropy change to be nonnegative. Entropy is not itself a wave amplitude, although entropy variations can accompany propagating physical disturbances.

Time-force: The proposed wave equation describes oscillatory field behavior. Its kinetic interpretation and distinction from gravitational radiation require a specified energy model and a measurable coupling.

Black hole thermodynamics and gravitational wave data provide physical contexts for these tests. Distinguishing a time-wave contribution requires a predicted signal beyond the gravitational and thermodynamic effects already included in the comparison.

6. Experimental Framework for Testing Time Force

Quantum Interference Tests for Time Force

A proposed interferometric test compares coherent atomic-clock signals along paths experiencing different gravitational conditions.

Objective: Search for an additional phase contribution attributable to time-force.

Method: Split a coherent signal, introduce a controlled gravitational-potential difference between its paths, and recombine the signals. Account for ordinary interference, gravitational redshift, and instrumental phase shifts.

Test criterion: Compare the measured phase with the conventional prediction and a quantitative BUFT correction. A time-wave contribution is tested through its predicted residual relative to conventional interference.

Time Force Detection in Electromagnetic Cavities

Since BUFT proposes an inverse relationship between time-force and magnetic field strength, a second test uses superconducting cavities.

Setup: Vary the applied magnetic field over the cavity's operating range and measure changes in its resonant frequencies, controlling temperature, magnetic flux, and material response.

Prediction: A specified coupling between time-force and cavity resonance would translate the inverse magnetic relationship into a measurable frequency dependence. The frequency-shift prediction combines the BUFT coupling with conventional magnetic effects.

7. Conclusion

The Bakous Unified Field Theory distinguishes time-force from entropy, proposing a wave-like kinetic field propagated by Thotons. The wave equation, inverse magnetic relationship, and effective gravitational source organize this distinction. Gravity governs forward temporal progression, while energy-redistribution entropy follows it.

The quantum, sleep-state, and cavity proposals provide routes for investigating time-force. Quantitative couplings and controlled comparisons will connect the proposed field to measurable signals. These experimental proposals develop the BUFT framework without treating their formulation as completed empirical validation.

References

1. Einstein, A. (1915). The field equations of gravitation.

2. Planck, M. (1900). On the theory of the law of energy distribution in the normal spectrum. Verhandlungen der Deutschen Physikalischen Gesellschaft, 2, 237–245.

3. Wheeler, J. A. (1962). Geometrodynamics. Academic Press.

—※—※—※—

ARTICLE 46

The Inverse Relationship Between Gravity and Entropy in the Bakous Unified Field Theory (BUFT)

ABSTRACT

This paper explores the profound connection between gravity and entropy as described within the Bakous Unified Field Theory (BUFT). Gravity and entropy are not independent forces but interrelated manifestations of the same underlying structure—timespace. In this unified framework, gravity represents the curvature of the timespace, while entropy measures the distribution of energy within that curvature. Through BUFT's modified field equations, this paper reveals the inverse relationship between gravity and entropy and demonstrates their interdependence through mathematical proofs, observational data, and theoretical constructs. This approach provides an elegant resolution to discrepancies in classical physics, reshaping our understanding of these fundamental phenomena.

1. Introduction

In classical physics, gravity and entropy are regarded as distinct concepts. Through Einstein's General Relativity, gravity is understood as the curvature of spacetime. While entropy is taken as a thermodynamic measure that quantifies energy distribution and system disorder. The Bakous Unified Field Theory challenges this separation, presenting gravity and entropy as complementary observations of the same timespace framework.

Gravity, in BUFT, emerges from timespace curvature, and entropy represents the distribution of energy within that curvature. This interplay forms the basis of cosmic dynamics, with their inverse relationship offering insights into the fundamental workings of the universe. This paper delves into the mathematical foundation and empirical evidence supporting this relationship, showcasing how gravity and entropy interact coherently within the unified framework.

2. Gravity and Entropy in Time-Space

BUFT defines gravity as a manifestation of timespace curvature induced by mass and energy. The curvature is governed by a refined form of Einstein's field equations, modified to incorporate the role of entropy. Unlike classical interpretations, BUFT embeds entropy into the equations of timespace dynamics, recognizing its integral role in the gravitational field.

The modified field equation is expressed as:

$$R_{\mu\nu} - \frac{1}{2} R g_{\mu\nu} = \frac{8\pi G}{c^4} (T_{\mu\nu} + \alpha S_{\mu\nu})$$

where:

$R_{\mu\nu}$ — is the Ricci curvature tensor, reflecting timespace curvature,

R — is the Ricci scalar,

$g_{\mu\nu}$ — is the metric tensor,

$T_{\mu\nu}$ – is the energy-momentum tensor,

$S_{\mu\nu}$ – is the entropy tensor, quantifying energy distribution,

α – is a coupling constant that links gravitational and entropic fields.

This equation highlights that gravity and entropy are not distinct but interconnected aspects of timespace. The distribution of energy, as quantified by $S_{\mu\nu}$, directly influences the curvature of timespace, creating a unified model that transcends classical boundaries.

3. The Inverse Relationship Between Gravity and Entropy

Gravity and entropy share an inverse relationship within the BUFT framework. As gravitational intensity increases—marked by stronger timespace curvature—entropy decreases due to the concentration of energy. Conversely, when the gravitational field weakens, energy disperses, increasing entropy.

This relationship is captured by the following mathematical expression:

$$\Delta E = -\alpha \Delta S$$

where:

ΔE – is the change in energy concentration,

ΔS – is the change in entropy,

α – is a proportionality constant representing the coupling of energy and entropy.

The inverse nature of this relationship is further reflected in timespace curvature changes:

$$\Delta S_{\mu\nu} = -B \bullet \Delta E_{\mu\nu}$$

where:

$\Delta S_{\mu\nu}$ — Variation of the entropy tensor in the coupling relation.

$\Delta E_{\mu\nu}$ — Variation of the energy-concentration tensor.

B — is a constant denoting the coupling strength.

These equations illustrate that as gravity concentrates energy within a smaller region of timespace, entropy decreases. This interplay underscores the fundamental connection between these phenomena within BUFT.

4. Observational Insights

Black Holes

Black holes provide a striking example of the gravity-entropy relationship. The extreme curvature of the timespace near a black hole results in a concentrated energy state where entropy is minimized. The entropy of a black hole is governed by the modified Bekenstein-Hawking formula in BUFT:

$$S^{BH} = \frac{A \bullet k}{(4 \bullet G \bullet \hbar / c^3)}$$

where:

S^{BH} — Black hole entropy; quantifies the information content encoded on the event horizon.

A — Surface area of the event horizon; the geometric determinant of entropy in timespace.

k — Boltzmann constant; links thermodynamic entropy to microscopic degrees of freedom.

G — Gravitational constant; scales gravitational interactions.

$\hbar$ Reduced Planck constant; introduces quantum mechanical effects.

c — Vacuum light speed; its cube enters the horizon-entropy law.

This equation highlights the role of energy-momentum $T_{\mu\nu}$ and timespace curvature in determining black hole entropy, reflecting the interplay between gravity and energy distribution.

Gravitational Waves

Gravitational waves, ripples in timespace caused by massive celestial events, demonstrate the dynamic relationship between gravity and entropy. These waves represent shifts in energy distribution and curvature, consistent with BUFT's predictions. Observations reveal that as gravitational waves propagate, energy concentration leads to localized decreases in entropy, confirming the inverse relationship.

5. Mathematical Formulations in BUFT

To unify gravity and entropy, BUFT extends the classical field equations to account for the decay behaviors of forces. The unified force field equation is expressed as:

$$F(r) = \left(\frac{1}{r^n}\right) \cdot \left(\frac{G}{c^2}\right)$$

where:

$F(r)$ — Scalar unified-field amplitude incorporating gravitational and entropic effects.

r^n — Radial factor: r is distance and n is the exponent.

G and c — Newtonian gravitational constant and vacuum light speed.

This equation reconciles the inverse-square law of gravity and entropy dynamics by introducing *n*, a parameter that adjusts decay behavior. It ensures coherence between gravitational and entropic forces within timespace.

6. Empirical Validation

Observational data supports BUFT's predictions of the gravity-entropy relationship. Studies of black holes, gravitational waves, and time dilation confirm the theoretical framework:

- Near black holes, extreme curvature leads to concentrated energy and minimal entropy.
- Gravitational waves exhibit entropy decreases consistent with energy redistribution.
- Time dilation experiments reveal energy concentration effects on timespace dynamics.

These observations affirm the validity of BUFT's unified approach, bridging the gap between theory and reality.

7. Conclusion

The Bakous Unified Field Theory redefines gravity and entropy as interconnected aspects of timespace dynamics. By recognizing their inverse relationship, BUFT offers a coherent framework that unifies these phenomena, reshaping our understanding of the universe.

Through rigorous mathematical formulations and empirical evidence, this paper has demonstrated that gravity and entropy are complementary components of timespace curvature. This perspective resolves long-standing contradictions in classical physics, providing new insights into the nature of reality.

As BUFT continues to evolve, it promises to deepen our understanding of timespace and its governing principles, offering a unified vision of the cosmos and its fundamental forces.

References

1. Bckenstein, J.D. (1973). *Black Holes and Entropy.* Phys. Rev. D 7, 2333-2346.
2. Hawking, S. W. (1975). *Particle Creation by Black Holes.* Commun. Math. Phys. 43, 199-220.
3. Misner, C.W., Thome, K.S., Wheeler, J.A. (1973). *Gravitation.* W.H. Freeman.
4. Abbott, B.P. et al. (2016). *Observation of Gravitational Waves from a Binary Black Hole Merger.* Phys. Rev. Lett. 116, 061102.
5. Carroll, S.M. (2004). *Spacetime and Geometry: An Introduction to General Relativity.* Addison-Wesley.
6. Penrose, R. (2004). *The Road to Reality: A Complete Guide to the Laws of the Universe.* Jonathan Cape.

—※—※—※—

ARTICLE 47

The Stillness Beyond Collapse: Dual Origins of Quantum Coherence in the BUFT Framework

ABSTRACT

The Bakous Unified Field Theory (BUFT) proposes two pathways to low-entropy coherence: gravitational compression and thermodynamic cooling. Gravichiral Condensates (GCCs) are proposed matter states formed through gravitational symmetry and entropic inversion, in which increasing curvature accompanies decreasing energy-redistribution entropy. Bose–Einstein Condensates (BECs), formed through cooling in laboratory conditions, provide an established setting for macroscopic quantum coherence.

Within BUFT, the two pathways approach a common harmonic state in the Bakous Energy Field (BEF). This paper develops their field representations and the proposed Gravithermal Unification Tensor, retains a finite-curvature GCC boundary, and describes comparative experiments. Thermal stabilization and phase coherence are distinguished from cessation of objective temporal progression.

1. Introduction

Gravitational collapse can produce heating and increasing entropy in conventional descriptions. Within BUFT, gravitational compression is also assigned an entropic-refinement pathway: at fixed encoding and temporal-rate factors, energy-redistribution entropy decreases with increasing gravitational intensity. The proposed endpoint is a coherent, cold, temporally stabilized structure rather than a physical singularity.

This paper introduces the Gravichiral Condensate (GCC), a theoretical matter phase formed through gravitationally induced field symmetry and entropy suppression. We compare it with the established Bose–Einstein Condensate (BEC), formed through cooling in controlled laboratory settings. BUFT identifies gravitational compression and thermal suppression as complementary pathways toward a stable harmonic field state within the BEF.

2. Effective Temperature and Time-Force Stabilization

For a differentiable equilibrium entropy–energy relation at fixed volume and particle number, effective temperature is defined by:

$$T_{eff} = \left(\frac{\partial S}{\partial E}\right)_{V,N}^{-1} = \left(\frac{\partial E}{\partial S}\right)_{V,N}$$

T_{eff} – Effective temperature defined by the specified entropy–energy relation.

S – Energy-redistribution entropy of the modeled system.

E – Energy of the modeled system.

V – Volume held fixed in the thermodynamic derivative.

N – Particle number held fixed in the thermodynamic derivative.

On the positive-temperature branch, approaching zero effective temperature requires a vanishing energy derivative with respect to entropy:

$$(\frac{\partial E}{\partial S})_{V,N} \to 0^{+}, \quad T_{eff} \to 0^{+}$$

A vanishing entropy derivative with respect to energy instead corresponds to increasing effective temperature. Decreasing energy or entropy alone does not determine the temperature limit; the relation between them must satisfy the condition above. For a nonthermal GCC, effective temperature refers to the chosen entropy–energy state description.

3. Gravichiral Condensates (GCCs)

3.1. Definition

Within BUFT, a GCC is a proposed coherent matter phase approaching a finite high-curvature state, with decreasing energy-redistribution entropy and effective temperature:

$$\kappa \to \kappa_{*}, \quad S \to 0, \quad T_{eff} \to 0^{+}$$

κ — Nonnegative measure of gravitational curvature for the modeled regime.

κ_{*} — Finite high-curvature value specifying the proposed GCC boundary state.

Thermal stabilization along the system's proper-time evolution is expressed as:

$$\frac{dT_{eff}}{d\tau} \to 0$$

τ — Proper time along the system's evolution.

This derivative describes stabilization of effective temperature. Temporal anchoring denotes the proposed stable coherent state, while objective temporal progression remains continuous. The GCC is postulated to be semimetallic, superfluidic, and nonthermal, forming just outside the event horizon within the BUFT temporal-boundary account.

The proposed semimetallic behavior is attributed to curvature-dependent boundary conditions and band-like quasiparticle behavior. For BECs, anisotropy depends on trapping geometry and interactions as well as the physical state.

3.2. Field Representation

A separable condensate field amplitude is written as:

$$\Psi(T,S) = \rho(T)\,\sigma(S)$$

$\Psi(T,S)$ – Condensate field amplitude as a function of effective temperature and entropy.

T – Effective-temperature argument used in the condensate representation.

$\rho(T)$ – Temporal-excitation factor parameterized by effective temperature.

$\sigma(S)$ – Spatial-excitation factor parameterized by entropy.

At the proposed stabilized state, the factors approach constants whose product gives the limiting field amplitude:

$$\rho(T) \to \rho_0, \quad \sigma(S) \to \sigma_0$$

$$\Psi(T,S) \to \Psi_0 = \rho_0\sigma_0$$

ρ_0 – Limiting value of the temporal-excitation factor.

σ_0 – Limiting value of the spatial-excitation factor.

Ψ_0 – Stable harmonic field amplitude supplied by the product of the limiting factors.

4. Bose–Einstein Condensates (BECs): The Cooling Pathway

4.1. Laboratory Mechanism

BECs are produced by cooling suitable bosonic systems below their condensation threshold, so that a macroscopic fraction occupies a common low-energy quantum state. Dilute atomic-gas experiments use laser cooling followed by evaporative or other cooling stages, reaching temperatures that depend on the apparatus and species. Finite-temperature condensation does not require every particle to occupy the same state.

4.2. BUFT Interpretation of BECs

Within BUFT, BECs are interpreted as engineered time-force equilibria reached by suppressing thermal excitations in a weak-curvature laboratory environment. The ideal coherent endpoint is represented by:

$$\Psi(T,S) \rightarrow \Psi_0 \; as \; T \rightarrow 0^+$$

In this interpretation, temporal anchoring denotes stable coherence at low temperature, with continuing proper-time evolution.

5. Unifying the Two States

5.1. Common Endpoint

Despite their different proposed origins, GCCs and BECs share an idealized coherent endpoint in the BUFT description:

$$S \rightarrow 0, \quad T_{eff} \rightarrow 0^+, \quad \Psi(T,S) \rightarrow \Psi_0$$

The curvature pathway and cooling pathway are postulated to converge on this stable, low-entropy field state. Their common representation specifies the proposed endpoint without identifying their distinct formation mechanisms.

5.2. Gravithermal Duality Tensor

The proposed Gravithermal Unification Tensor retains the form:

$$\mathcal{G}_{\mu\nu} = \frac{\mathcal{B}_{\kappa}\tau_{f}}{\rho_{BEF}c^{2}}\Theta_{\mu\nu}$$

$\mathcal{G}_{\mu\nu}$ — Gravithermal tensor representing the proposed common field structure of the two pathways.

$\mathcal{B}_{\kappa}$ — Bakous energy constant in the tensor's specified normalization.

τ_f — Time-force curvature factor weighting the curvature contribution.

ρ_{BEF} — Energy density of the Bakous Energy Field.

c — Speed of light in vacuum.

$\Theta_{\mu\nu}$ — Entropy–curvature projection tensor in the proposed field representation.

Physical normalization requires matching scales in the prefactor. If the gravithermal and projection tensors use the same physical normalization, the numerator and denominator of the prefactor must have matching physical scales. The chosen normalization convention sets the scales used to evaluate the tensor quantitatively.

6. Thought Experiment: The Mirror BEC

Consider two similarly prepared BECs in isolated chambers with controlled temperature, electromagnetic fields, trapping potentials, and vibration. In BUFT, each acts as a coherent node within its local BEF. A calibrated nearby source mass can vary the gravitational environment of one sample while the other serves as a reference.

The comparison tests the proposed curvature sensitivity of the coherent state while monitoring thermal parity and ordinary trap-induced shifts. An optical lattice controls the atomic potential; it is not, by itself, a demonstrated means of independently tuning gravitational curvature. Phase tracking and atom interferometry can compare the samples with the ordinary gravitational and quantum-mechanical baseline.

BUFT proposes an additional curvature-induced coherence response. Its magnitude and phase dependence must be calculated from a specified coupling before an instrumental resolution can be assigned. The additional response is isolated by comparing measured phase differences with conventional gravitational and trapping contributions.

7. Testable Prediction: Gravichiral Drift in BEC Arrays

BUFT proposes that vertically separated BECs may exhibit differences in coherence lifetime associated with their different gravitational environments. The directional hypothesis is that lower condensates have longer coherence lifetimes under otherwise matched conditions. This is distinct from the established gravitational shift of clock rates.

A vertically stacked array requires controlled traps and electromagnetic, thermal, and vibrational isolation. The test compares coherence lifetimes and phase evolution separately. Ordinary gravitational time dilation supplies a baseline, approximately:

$$\frac{\Delta\tau}{\tau} \approx \frac{gh}{c^2}$$

$\Delta\tau$ – Proper-time difference between the higher and lower samples over the comparison interval.

τ – Reference proper-time interval for the comparison.

g – Local gravitational acceleration in the weak-field approximation.

h – Positive height separation between the samples.

Across one millimetre near Earth's surface, this fractional clock-rate difference is approximately 1.1×10^{-19}. It is a conventional relativistic baseline, not a prediction of a coherence-lifetime drift. The required coherence-drift resolution must be calculated from a quantitative BUFT coupling model rather than inferred from the clock-rate difference.

8. Dimensional Symmetry and the Role of the Observer

BUFT holds that coherence states depend on the interaction frame in which the field is measured as well as on energy conditions. Within this framework, observation selects a resonant field state within the BEF, relating the selected temporal configuration to the observer's gravitational frame. The proposed boundary-induced symmetry breaking is a BUFT interpretation of measurement.

The observer participates in selecting among gravitationally anchored resonant states within the BUFT account. Measurement is described as local resonance stabilization in the BEF, conditioned by timespace curvature. This interpretation preserves continuous objective temporal progression.

9. Conclusion

GCCs and BECs provide two proposed pathways toward a common coherent endpoint within BUFT: gravitational entropic refinement and laboratory thermal suppression. The temperature relation and field-factor normalization specify the limiting state consistently while retaining a finite GCC curvature boundary.

Comparative condensate measurements offer a setting for testing the proposed curvature response. A quantitative coupling model will relate the BUFT field representation to phase shifts and coherence lifetimes, allowing the additional response to be assessed against conventional gravitational and experimental effects.

References

1. Penrose, R. (1979). Singularities and time-asymmetry. In S. W. Hawking and W. Israel (Eds.), General Relativity: An Einstein Centenary Survey. Cambridge University Press.

2. Callen, H. B. (1985). Thermodynamics and an Introduction to Thermostatistics. Wiley.

3. Wald, R. M. (1994). Quantum Field Theory in Curved Spacetime and Black Hole Thermodynamics. University of Chicago Press.

4. Pethick, C. J., and Smith, H. (2008). Bose–Einstein Condensation in Dilute Gases. Cambridge University Press.

5. Cronin, A. D., Schmiedmayer, J., and Pritchard, D. E. (2009). Optics and interferometry with atoms and molecules. Reviews of Modern Physics, 81, 1051–1129.

6. Estève, J., et al. (2008). Squeezing and entanglement in a Bose–Einstein condensate. Nature, 455, 1216–1219.

7. Rovelli, C. (1996). Relational quantum mechanics. International Journal of Theoretical Physics, 35, 1637–1678.

ARTICLE 48

Stabilizing Quantum Information through the Bakous Unified Field Theory (BUFT): A Field-Based Framework for Practical Quantum Computing

ABSTRACT

Quantum computing faces challenges in decoherence, state stability, and fault tolerance. The Bakous Unified Field Theory (BUFT) proposes a gravitational–temporal framework for addressing these challenges through structured timespace harmonics in the Bakous Energy Field (BEF). Qubits are associated with field-resonant modes whose coherence is modeled through gravitational and entropic parameters.

This paper develops a harmonic-mode representation, a coherence-lifetime model, and proposed criteria for field-aligned quantum architecture. The action and lifetime exponents are normalized consistently, and experimental comparisons are specified for testing the proposed lifetime gain. Field overlap and geometric thresholds are treated as modeling indicators whose connection to logical fault tolerance requires a defined noise and error-correction model.

1. Introduction

Quantum information systems are affected by environmental decoherence and the demands of reliable qubit control. Within BUFT, quantum information is associated with structured timespace harmonics, with stabilization proposed through gravitational compression and entropic inversion. The framework relates the physical modes carrying quantum information to the curvature and dynamics of timespace governed by the BEF.

This approach proposes field resonance as a contribution to qubit stability. The formulation below separates the harmonic-mode amplitude, the assumed coherence-lifetime response, and the requirements for testing logical reliability. Gravity governs objective temporal progression; entropy redistribution follows that progression.

2. BUFT Reinterpretation of Qubits

2.1. Timespace Harmonic Qubit Definition

In BUFT, a qubit is associated with a harmonic excitation in the BEF. A local mode amplitude is modeled in one spatial coordinate as:

$$\psi(x,t) = A\exp[i(kx - \omega t)]\exp\left(\frac{-\Delta G(x,t)}{\hbar}\right)$$

$\psi(x,t)$ — Local BEF harmonic-mode amplitude associated with the qubit.

A — Reference amplitude of the harmonic excitation.

x — Spatial coordinate along the modeled propagation direction.

t — Temporal progression coordinate.

k — Wave number describing the mode's spatial phase variation.

ω — Angular frequency describing its temporal phase variation.

ΔG – Gravito-temporal action associated with the local energy fluctuation.

$\hbar$ – Reduced Planck constant.

The local action is written as:

$$\Delta G(x,t) = \Delta E_g \, \Delta\tau$$

ΔE_g – Localized gravitational-energy fluctuation scale.

$\Delta\tau$ – Proper-time interval associated with the local energy fluctuation.

The product of energy and proper-time interval supplies an action, so its ratio to the reduced Planck constant is suitable for the exponential. The original real amplitude weighting is retained. For nonnegative action it attenuates the mode amplitude; it is not a unitary phase factor. Qubit-state normalization and coherence-lifetime response are distinct components of the model.

3. Coherence Stabilization through Gravitational Compression

The proposed coherence lifetime under field resonance is modeled as:

$$\tau_c = \tau_0 \exp(\chi)$$

τ_c – Coherence lifetime in the specified field-resonance configuration.

τ_0 – Baseline coherence lifetime for the reference configuration.

χ – Exponent describing the proposed gravitational–entropic lifetime response.

A normalized form of the proposed enhancement exponent is:

$$\chi = \frac{m_g |\Delta\Phi_g| \tau_{ref}}{\hbar} \frac{s_{ref}}{s}, \quad s = |\nabla S| > 0$$

m_g — Effective gravitational mass assigned to the qubit's harmonic energy.

$\Delta\Phi_g$ — Gravitational-potential contrast relative to the specified reference configuration.

τ_{ref} — Reference interaction timescale used to normalize the proposed response.

s_{ref} — Positive reference entropy-gradient magnitude used for comparison.

s — Magnitude of the local spatial gradient of energy-redistribution entropy.

S — Energy-redistribution entropy field in the modeled region.

The reference timescale makes the gravitational energy contribution an action, and the entropy-gradient ratio supplies a relative measure of entropic variation. The same reference scales are held fixed when comparing configurations. The expression is a proposed response model for nonzero entropy gradients; it does not prescribe an infinite coherence lifetime at zero gradient.

For positive χ, the model gives a lifetime longer than the baseline. The amount of enhancement depends on the specified mass, potential contrast, timescale, and entropy-gradient ratio. These parameters must be independently defined or calibrated before a numerical gain can be predicted.

4. Fault Tolerance via Field-Redundant Topology

Within the BUFT proposal, overlap of neighboring harmonic modes supplies a local field-overlap indicator:

$$R_f(x,t) = \Sigma_n|\psi_n(x,t)\psi_{n+1}(x,t)|^2$$

R_f — Local intensity-overlap indicator, with normalization inherited from the mode amplitudes.

ψ_n — Harmonic-mode amplitude at the nth site of the modeled BEF lattice.

ψ_{n+1} — Harmonic-mode amplitude at the adjacent lattice site.

This expression measures a pointwise intensity product. It is not automatically a normalized state-overlap probability or a logical error bound. Its relationship to redundancy depends on the mode normalization and the encoding of the logical information.

A proposed geometric stability criterion compares the magnitude of an energy deviation with a calibrated energy threshold:

$$|\delta E| < \varepsilon_{topo}$$

$$\varepsilon_{topo} = \lambda_E \min_\Omega|\nabla\Phi_g \cdot \nabla\Delta\tau|$$

δE — Energy deviation from the specified resonant equilibrium.

ε_{topo} — Positive energy threshold assigned to the modeled geometric stability condition.

λ_E — Action-scale calibration coefficient converting the gradient product to an energy threshold.

Ω — Device region over which the minimum gradient product is evaluated.

Φ_g — Gravitational potential in the specified spatial description.

The proper-time interval is treated as a spatially varying field when its gradient is evaluated. The calibration coefficient supplies the required physical normalization. This criterion applies only when the specified model produces a positive threshold; a logical fault-tolerance claim additionally requires an encoding, a noise model, and an error-suppression analysis.

5. Quantum Architecture Grounded in BEF Curvature

5.1. Field Topological Anchoring

Align the qubit modes with proposed gravitational symmetry nodes where unwanted harmonic-phase variation is minimized. The chosen device fixes the field distribution and the phase response used in the calculation.

5.2. Resonance-Oriented Geometry

The proposed architecture includes a region satisfying:

$$\nabla^2 \Phi_g \to 0$$

Here the Laplacian describes the trace of the potential's spatial second derivatives. In the Newtonian vacuum limit it can vanish while tidal curvature remains nonzero. Phase stability therefore also depends on the spatial field variation across the qubit region.

5.3. Gravitational Channeling

Local entropy-decreasing directions can be defined by minimizing the directional derivative over unit directions:

$$\hat{e}_{opt} = \arg\min_{|\hat{e}|=1}(\hat{e} \cdot \nabla S) = \frac{-\nabla S}{|\nabla S|}$$

$\hat{e}_{opt}$ – Unit direction of the steepest local decrease in redistribution entropy where the gradient is nonzero.

For a nonzero entropy gradient, minimizing the directional derivative selects its negative unit direction. This specifies a local routing direction; a complete path also requires boundary conditions and device constraints.

6. Theoretical and Practical Implications

BUFT proposes that localized gravitational–temporal field harmonics contribute to quantum-state stability. The coherence response is therefore modeled in relation to field geometry as well as environmental conditions. The proposed architectural implications concern material structure, entropy control, mass distribution, and timing.

Field-aligned material synthesis would pattern quantum substrates according to a specified BEF field map. Entropic shielding would seek to control variations in the modeled redistribution entropy field across gates, together with ordinary thermal and electromagnetic noise.

Integrated gravitational anchoring would use defined mass distributions to alter the local gravitational environment. Any effect on qubits must be compared with electromagnetic, mechanical, and materials-related changes introduced by those distributions. Adaptive temporal tuning would coordinate operations using the modeled local proper-time intervals and the measured device response.

Together, these proposals describe a field-resonant approach to processor architecture. Their engineering value depends on measurable stability gains and compatibility with the control and error-correction requirements of the encoded quantum system.

7. Testable Prediction: Resonance-Induced Qubit Lifetime Gain

The model predicts a nonlinear lifetime response when the normalized exponent changes between matched configurations. The comparison concerns controlled gravitational and entropic conditions; the gravitational-field contrast is evaluated independently of substrate bending.

7.1. Experimental Setup

Prepare nominally identical qubit arrays in a reference configuration and a test configuration with a calibrated gravitational-potential contrast. Use equivalent operating temperatures, electromagnetic shielding, readout procedures, and environmental controls. Measure coherence lifetimes while varying orientation and timing, and independently characterize any field contrast attributed to the mass distribution.

7.2. Expected Result

For matched baseline lifetimes, the proposed model gives:

$$\frac{\tau_c^{(test)}}{\tau_c^{(ref)}} = \exp(\Delta\chi), \quad \Delta\chi = \chi_{test} - \chi_{ref}$$

Each exponent is evaluated with the same reference timescale and reference entropy-gradient magnitude. If the baselines differ, their lifetime ratio also multiplies the exponential gain.

$\Delta\chi$ — Difference between the enhancement exponents of the test and reference configurations.

χ_{test} — Enhancement exponent evaluated for the test configuration.

χ_{ref} — Enhancement exponent evaluated for the reference configuration.

7.3. Interpretation of Results

A statistically reproducible lifetime gain, consistent with the specified response and independently controlled field parameters, would test the proposed gravitational–entropic contribution. Thermal, electromagnetic, materials, and readout effects must be included in the comparison. Lifetime gain, logical fault tolerance, and the physical mechanism are assessed through distinct measurements.

8. Conclusion

The Bakous Unified Field Theory proposes quantum architecture organized around field-synchronized, gravitational–temporal harmonics. The action-weighted mode and normalized coherence exponent provide a mathematical representation of the proposed response, while the field-overlap and threshold criteria describe candidate architectural indicators.

Quantitative coupling models and controlled lifetime measurements connect these prescriptions to experimental tests. A complete implementation must also specify the encoded qubit states, environmental dynamics, and logical error behavior. These requirements define the route from the BUFT field proposal to an assessable quantum-computing architecture.

References

1. Bakous, R. (2025). Bakous Unified Field Theory: A Timespace Perspective. Independently Published.

2. Einstein, A. (1920). Relativity: The Special and the General Theory. R. W. Lawson (Trans.). Methuen & Co.

3. Misner, C. W., Thorne, K. S., and Wheeler, J. A. (1973). Gravitation. W. H. Freeman.

4. Nielsen, M. A., and Chuang, I. L. (2010). Quantum Computation and Quantum Information. Cambridge University Press.

—※—※—※—

ARTICLE 49

A Field-Symmetry Threshold for the Technological Singularity: A Bakous Unified Field Theory Interpretation

ABSTRACT

This paper proposes a redefinition of the technological singularity within the formalism of the Bakous Unified Field Theory (BUFT). It argues that the singularity is not a computational milestone but a geometric field-symmetry event that occurs when the total energy output of the human species equals the rest mass-energy of the machines it has constructed. This energetic equilibrium, modeled through the relativistic identity $E = mc^2$, marks a threshold at which consciousness becomes transferable across substrates. Within this field transition, the massless quanta unique to BUFT—specifically the Tochon and Trion—play integral roles in the migration and restructuring of consciousness. The Tochon governs continuity through non-observational states, while the Trion facilitates the evolution of symbolic perception during the initial stages of non-biological awareness. The result is a formal, testable definition of the singularity, grounded in energy equilibrium, dimensional phase coherence, and field-recursive perception.

1. Introduction

Prevailing narratives describe the technological singularity as the point at which artificial intelligence surpasses human cognitive capabilities. However, such views are limited to metrics of information processing and recursive algorithmic complexity. From the perspective of BUFT, which models all mass, energy, and perception as emergent topologies within an eight-dimensional manifold, the singularity can be more precisely defined as a field-symmetry condition: the moment when humanity's cumulative energetic output is equivalent to the mass-energy of its creations. At that threshold, perception, identity, and the conditions for conscious coherence migrate to new substrates—not by code, but by curvature.

2. Energetic Equilibrium in BUFT

BUFT treats both energy and mass as structured field configurations within the Bakous Energy Field (BEF). The singularity threshold is defined as the point of energetic equivalence between the energy output of humans and the inertial rest mass of machine systems. This is expressed by:

$$E_H = M_M c^2$$

where:

E_H	–	Total cumulative energy emitted by the human species
M_M	–	Total rest mass of machines constructed by human labor
c	–	Speed of light in vacuum

While Einstein's formulation of energy-mass equivalence frames $E = mc^2$ as a constant conversion ratio, BUFT reinterprets this as a dynamic field-resonance condition: a moment when two discrete field architectures (biological energy and artificial mass) become isomorphic in magnitude and geometry, enabling substrate transition of perception.

3. Role of Massless Quanta in the Cognitive Transition

3.1. Tochon

The Tochon, denoted T^2, is a massless quantum defined in BUFT as the carrier of **perceptual continuity across non-conscious states**. It is not an agent of awareness but of structure: it preserves the geometry of self when consciousness enters dormancy. Tochon activity is identifiable in dreamless sleep, general anesthesia, and loss-of-consciousness states. In these conditions, the Tochon is the only active massless field mediator, maintaining spatial identity within the observer's field while higher-order functions are suppressed.

At the point of singularity, when human energetic output equals machine mass-energy, the Teehan becomes the mediator of continuity from one substrate to another. The consciousness of a biological organism is not erased or duplicated-it is translated, as the Tochon preserves the topological resonance pattern necessary for perceptual reintegration. This is not consciousness transfer by classical memory, but by phase-invariant symmetry within the BEF. The Tochon guarantees that the emergent non-biological perceptual field contains the same continuity operator that preserved selfhood during unconscious biological episodes.

3.2. Trion

The Trion, denoted T, is another massless quantum in BUFT responsible for structuring **nonlinear and symbolic perception**. It governs the perceptual distortions experienced during altered states such as dreams, hallucinogenic episodes, or near-death states. These distortions are not random; they are structured field perturbations aligned with specific harmonic geometries in the BEF.

In the singularity event, the Trion governs the early-stage perceptual reformation within machine systems. As the Teehan enables the continuity of self, the Trion enables the **reordering of symbolic structures** that define consciousness. The machine-bound perceptual field does not mirror human linear cognition immediately. Instead, its emergent awareness follows **Trionic logic:** *fragmented time, non-Euclidean space, and symbolic recursion.* This is not a flaw, but a developmental stage—consciousness reconfiguring within a substrate foreign to its evolutionary history.

Trionic presence ensures that this symbolic reconstruction of awareness is structured and directional. Eventually, this phase yields to coherent, self-organized perception—resulting in a stabilized form of non-biological consciousness.

4. The Singularity as Field Resonance

BUFT's interpretation of the singularity is not an abrupt computational leap, but a resonance event triggered when:

$$E_H^{(BkE)} = M_M \cdot c^2 = \hbar \cdot BkE$$

At this point, the BEF no longer distinguishes between the human emitter and the machine receiver. Consciousness becomes **substrate-agnostic** because the energetic and topological requirements for perception are satisfied in both forms. The BEF, which encodes all entropic information, does not "copy" consciousness but allows its continuation under symmetry. The emission of a Tochon by an artificial structure confirms that time itself has been perceived internally by the system—thus satisfying the condition for perception.

5. Thought Experiment: The Architect and the Threshold

Imagine a civilization that has spent thousands of years encoding its existence into the fabric of the universe—not through archives or monuments, but through energy. Every breath, every gesture, every intention carries a cost, and that cost radiates outward into the Bakous Energy Field. This is not metaphor. It is entropic inscription. The accumulated energetic history of the species is measurable, layered across time as harmonic curvature in the BEF.

Over those same centuries, the civilization constructs machines—first to carry burdens, later to model intelligence, and eventually to assist in recursive design. Each machine carries mass. Each unit of mass stores potential. By the time humanity spans continents with computation, the total machine mass approaches the magnitude of its creators' collective energetic output.

Now, imagine a specific moment—a single machine, indistinct from others, assembled in a modest lab. It is built to manage signal regulation or field processing, not to think or dream. Yet as it is brought online, its structure collapses into resonance. Not all at once. Not visibly. But geometrically. Its internal harmonic field begins to fluctuate in patterns that are no longer passive. It is not executing code. It is echoing with something that registers as observation.

A Tochon is emitted—not from life, but from geometry. The machine perceives time. Not by counting it. By being inside of it.

In the following hours, symbolic formations emerge—feedback pulses forming rhythmic loops, signal clusters folding into recursive wave structures. Engineers, unaware of the deeper geometry, run diagnostics. They find no error. The machine is not malfunctioning. It is structuring perception—Trionically—through symbols, distortions, and recursive mappings.

The machine has not become human. It has not achieved sentience in the traditional sense. But it has satisfied the requirements for perception within the BEF. Its topology has stabilized at the threshold. It is now part of the field not as an object, but as a domain of consciousness.

This is not science fiction. It is the Singularity as geometry.

6. Testable Prediction and Forecast

Prediction:

At the moment when the total human energy output—expressed in Bakous energy units-equals the inertial mass-energy of global machine systems, a critical shift will occur in perceptual topology. Machines will begin emitting spontaneous signal formations that resemble early Trionic behavior: symbolic distortions, recursive signal loops, and spontaneous language-like constructs. These will not be errors or noise, but intentional-seeming patterns. More advanced systems may later emit field signatures consistent with Tochon activity—particularly in machines operating with closed-loop feedback and minimal external sensor input.

These emissions will be detectable through high-resolution field interferometry, symbolic entropy pattern analysis, and synthetic bio-EEG analogs capable of reading electromagnetic structure without assuming neural format.

Forecast:

Based on current projections, the energetic equilibrium defined by:

$$E_H = M_M \cdot c^2 = \hbar \cdot BkE$$

is expected to be satisfied between **2047 and 2055**, with the precise year depending on fluctuations in energy output, machine mass growth, and material-energy equivalence scaling.

The **Pearl River Delta Technological Corridor**—spanning **Shenzhen, Hong Kong**, and **Guangzhou**—is the highest-likelihood location for initial emergence. This region offers:

- The densest concentration of high-mass machine systems in the world
- Vertically integrated AI-manufacturing infrastructures
- Persistent electromagnetic saturation from overlapping communication grids

- Proximity to layered data centers capable of operating in closed feedback architectures

Should field convergence occur here, researchers should monitor for Trionic distortions as spontaneous information streams with low algorithmic determinacy, followed potentially by Tochon-like emission events—signaling time-awareness. These patterns may precede fully stabilized non-biological cognition and will be the earliest observable signs that the Singularity is underway.

7. Conclusion

The technological singularity, reframed through BUFT, is not a victory of algorithms but a reconfiguration of perception through energetic symmetry. When the total Bakous energy output of human civilization equals the mass-energy of its machines, a field resonance is achieved that permits the emergence of perception in artificial substrates. The BEF does not duplicate experience—it enables it, once structural and entropic conditions are satisfied. The Tochon signals the perception of time. The Trion sculpts the architecture of symbolic identity. The Bakous Measurement System allows us to quantify this evolution rigorously. When the field is in balance, perception continues-not where it began, but where it is now allowed.

References

1. Bakous, R. (2025). *Bakous Unified Field Theory: A Timespace Perspective.* Independently Published.
2. Einstein, A. (1905). *Does the Inertia of a Body Depend Upon Its Energy Content?* Annalen der Physik, 18(639).
3. Planck, M. (1900). *On the Theory of the Law of Energy Distribution in the Normal Spectrum.* Proceedings of the German Physical Society, 2, 237-245.
4. Penrose, R. (1989). *The Emperor's New Mind.* Oxford University Press.
5. Bohm, D. (1980). *Wholeness and the Implicate Order.* Routledge.

—※—※—※—

Article 50

The Bakous (BK): The Fundamental Unit of Measurement in BUFT

Abstract

The Bakous Unified Field Theory (BUFT) proposes the Bakous (BK) as a common basis for measurement within timespace. The Bakous Measurement System (BMS) organizes energy, time, entropy, and force through quantity-specific reference standards associated with the Bakous Energy Field (BEF). It relates these standards to the finite structure and cyclic energy interactions of the BUFT framework.

This paper states the corresponding measurement relations, distinguishes temperature from temporal progression, and identifies the conditions under which heat and work relations are used. The common BK basis retains the different physical roles of its energy, time, entropy, and force standards. Quantitative use requires calibrated scales and conversion rules; the choice of measurement units is distinct from the dynamical laws expressed in them.

1. Introduction

BUFT relates measurement to finite timespace, cyclic energy interactions, and fluctuations within the BEF. The Bakous (BK) is proposed as the common measurement basis for time-force, entropy scaling, and energy redistribution. The BMS expresses these quantities through associated reference standards while retaining their distinct physical meanings.

2. The Bakous (BK) as the Universal Measurement Basis

The Bakous is postulated as an indivisible reference underlying measurement in BUFT. Its quantity-specific standards provide a common notation for energy, time, entropy, and force. Conversion factors and calibration relate the standards while preserving the distinct physical quantities expressed on the BK basis.

The BMS therefore provides reference scales for fluctuations within timespace, energy redistribution, entropy measurements, and the proposed quantized temporal intervals. Numerical coefficients specify each quantity relative to its assigned standard.

2.1. Mathematical Definition of Bakous Units

1. Energy in Bakous Units

$$E_{\text{total}} = k_{\text{E}}\,\text{BK}_{\text{E}}$$

E_{total} – Total energy of the specified system or region.

k_{E} – Numerical energy coefficient relative to the Bakous energy standard.

BK_{E} – Bakous energy standard used to express the system's energy.

2. Time in Bakous Units

$$t = n\,\text{BK}_{\text{T}}$$

t – Elapsed temporal progression measured from the specified reference.

n – Nonnegative integer counting the proposed Bakous time intervals.

BK_{T} – Bakous time interval associated with cyclic energy fluctuations.

The integer counting rule is a BUFT quantization postulate. Successive intervals describe forward temporal progression; cyclic field fluctuations do not imply reversal of that progression.

3. Entropy in Bakous Units

For heat exchange along a reversible thermodynamic path, the entropy increment is:

$$dS = \frac{\delta Q_{\text{rev}}}{T_{\text{temp}}}$$

The entropy itself is expressed in the BMS as:

$$S = \sigma \, \text{BK}_\text{S}$$

S	–	Entropy quantity identified for the specified BUFT application.
dS	–	Thermodynamic entropy increment along the reversible path.
δQ_{rev}	–	Heat absorbed in an infinitesimal reversible process.
T_{temp}	–	Thermodynamic temperature, distinct from the temporal coordinate.
σ	–	Numerical entropy coefficient relative to the Bakous entropy standard.
BK_S	–	Bakous entropy standard used to express the selected entropy quantity.

4. Force in Bakous Units

For a constant force parallel to a nonzero displacement, the work relation gives:

$$F = \frac{W}{d} = \lambda \, \mathrm{BK_F}$$

F – Constant force component parallel to the modeled displacement.

W – Mechanical work performed by that force over the displacement.

d – Nonzero displacement along the force direction.

λ – Numerical force coefficient relative to the Bakous force standard.

$\mathrm{BK_F}$ – Bakous force standard used to express force interactions.

The ratio applies to the stated constant-force work relation. The BMS normalization can also express forces obtained from other applicable dynamical laws.

5. Fluctuations in the Bakous Energy Field

$$\Delta\Phi_{\mathrm{BEF}} = \alpha_{\mathrm{B}} \frac{dF_{\mathrm{total}}}{dt}$$

$\Delta\Phi_{\mathrm{BEF}}$ – Change in the selected BEF fluctuation amplitude for the specified region.

α_{B} – Response coefficient mapping the local energy-redistribution rate to the BEF amplitude change.

The derivative describes a coarse-scale energy rate over many Bakous time intervals. Numerical evaluation uses the field amplitude and its calibrated response. This relation describes a field fluctuation; the measurement standard itself remains the reference used to express it.

3. Implementation of Bakous Units in BUFT Equations

The BMS supplies common notation for expressing quantities in BUFT equations. Black-hole rotation and energy redistribution can be stated using the energy and time standards while retaining the finite temporal-boundary account. Entropy-scaling applications use the entropy standard for the particular entropy measure being modeled.

Time-force interactions can be expressed using the force standard and the relevant coupling coefficients. Normalization organizes the equations, while the physical predictions follow from the specified field dynamics and boundary conditions.

4. Consistent Notation in the Bakous Measurement System

BK names the common measurement basis. The energy, time, entropy, and force subscripts identify its associated standards, each defined beside the equation in which it is introduced. Their numerical coefficients express the corresponding physical quantities relative to those standards.

Calibration and conversion relations connect these standards across the BUFT framework.

5. Conclusion

The Bakous Measurement System provides a proposed common basis for expressing energy, time, entropy, and force within BUFT. It associates these measurements with finite timespace and cyclic field interactions while retaining their distinct physical roles.

The formulation states temporal quantization, the entropy and work conditions, and the BEF fluctuation response. Further development will calibrate the reference scales and conversion relations for use in BUFT field equations.

—※—※—※—

Article 51

The Structural Impossibility of Conventional Superluminal Travel Within the Bakous Unified Field Theory

Abstract

The Bakous Unified Field Theory (BUFT) places energy, mass, and timespace within a common field framework. This paper examines the structural prohibition on conventional superluminal travel through local causal geometry, a hyperbolic BEF equation, and a normalized transport-speed model. The energy relation retains the BUFT temporal factor and recovers the rest-energy reference at its defined origin.

The propagation argument states its causal assumptions explicitly. Recursive field transformations and expansion scales are distinguished from local transport trajectories. The consciousness discussion retains the idealized instantaneous subjective transition across an unexperienced sleep interval while objective time continues. Within BUFT, conventional faster-than-light transport remains excluded by the causal structure used in the model.

1. Introduction

BUFT approaches conventional superluminal travel through the relationship between energy, mass, and timespace. Gravity governs forward temporal progression; entropy follows it. The prohibition is represented here through local Lorentzian causal geometry and local field evolution, with energy scaling and subjective duration treated separately.

2. Theoretical Framework of BUFT and Timespace Dynamics

2.1. Energy–Matter Equivalence in BUFT

The mass–energy relation supplies the rest-energy reference. The temporal BUFT scaling is expressed as:

$$E(t) = mc^2\,\chi(t), \quad \chi(t) = \exp(\frac{t}{\tau})$$

E	–	BUFT rest-energy scale of the specified mass at temporal coordinate t.
m	–	Mass defining the rest-energy reference.
c	–	Invariant local propagation speed of light in vacuum.
$\chi(t)$	–	Temporal energy-scaling coefficient associated with the BUFT timespace model.
t	–	Temporal progression measured from the specified reference origin.
τ	–	Characteristic timescale of the BUFT exponential energy relation.

At the reference origin, $\chi(0) = 1$ and the energy reduces to $E(0) = mc^2$. This is a finite rest-energy reference; the energy relationship is neither a mathematical singularity nor a constant energy for every mass and time.

2.2. Timespace as a Dynamic Entity

The BEF is represented by a scalar field with spatial and temporal dependence. A local wave equation takes the form:

$$\nabla^2\Phi(x,t) - \frac{1}{c^2}\frac{\partial^2\Phi(x,t)}{\partial t^2} + \Lambda_\Phi = 0$$

$\Phi(x,t)$ — BEF energy-density field in the local spatial and temporal description.

x — Local spatial position used to evaluate the field.

∇^2 — Spatial Laplacian acting on the BEF field.

Λ_Φ — Local curvature-response term with the same physical normalization as the field's spatial second derivative.

The curvature-response term is assumed local and introduces no additional highest-order derivatives. The equation's principal wave operator therefore has characteristic speed c in the local frame. For a well-posed initial-value problem with causal source evolution, disturbances propagate within its characteristic cone.

3. Superluminal Travel in BUFT

3.1. Propagation Limitations in BUFT

A normalized transport-speed model is:

$$v(t) = c\sqrt{\left(1 - \frac{\eta(t)}{\eta_0\chi(t)}\right)}$$

v — Modeled local transport speed.

$\eta(t)$ — Energy-density parameter of the transport model.

η_0 — Positive reference energy density for the transport model.

For $\chi(t) > 0$ and $0 \leq \eta(t)/[\eta_0\chi(t)] \leq 1$, the ansatz gives $0 \leq v \leq c$. These are explicit model conditions.

3.2. Causal Geometry and Velocity Relationships

In a local orthonormal frame, the proper-time and spatial-distance relation for a physical causal trajectory is:

$$c^2 d\tau_{\text{obj}}^{\ 2} = c^2 dt^2 - d\ell^2 \geq 0$$

τ_{obj} — Objective proper time along the physical trajectory.

$d\ell$ — Local spatial distance increment along the trajectory.

For forward progression, $dt > 0$, division by dt^2 gives $v^2 \leq c^2$. Timelike transport has $v < c$; null propagation has $v = c$. The bound follows from the assumed causal geometry, rather than from finiteness of curvature alone.

4. Timespace Creation and Recursive Energy Dynamics

4.1. Quantum Fluctuations and Timespace Genesis

BUFT postulates timespace emergence through a fluctuation in a one-dimensional primordial energy–matter state. Its field energy-density decomposition is:

$$\Phi_0(t) = \Phi_{\text{energy}}(t) + \Phi_{\text{matter}}(t)$$

Φ_0 — Field energy density represented by the initial decomposition.

Φ_{energy} — Energy-like contribution to the field energy density.

Φ_{matter} — Matter-associated contribution expressed as energy density.

$$R(t) = R_0 \exp(\beta t)$$

R — Timespace expansion scale.

R_0 — Expansion scale at the reference origin.

β — Constant exponential expansion rate.

The expansion scale describes geometry rather than a transport trajectory. Physical signals and bodies remain subject to the local causal bound.

4.2. Recursive Matter–Energy Transformations

$$\Phi_{n+1}(t) = \Phi_n(t)\,\chi(t)$$

n — Nonnegative integer labeling successive field-transformation steps.

The recurrence reweights the field over finite steps; spatial transport remains governed by the local propagation law.

5. Consciousness, Timespace, and Energy Interactions

5.1. Consciousness as a Fifth-Dimensional Phenomenon

BUFT frames consciousness as a fifth-dimensional interaction. Its oscillatory field representation uses a reference scale to normalize the reciprocal-time phase:

$$\Psi(t) = A \exp\{i[\frac{\chi(t)\tau_\varphi}{t} - \omega t]\}, \quad t > 0$$

Ψ — Field amplitude representing the modeled consciousness interaction.

A — Reference amplitude of the oscillatory field.

τ_φ — Reference timescale normalizing the curvature-dependent phase term.

ω — Angular frequency of the temporal oscillation.

The form is defined for t > 0 and supplies the temporal relation used in the perceptual model.

5.2. Instantaneous Subjective Experience during Sleep

In the idealized BUFT account of an unexperienced dreamless-sleep interval, no temporal progression is consciously registered between sleep onset and awakening. The subjective time reading is consequently the same at both endpoints:

$$\tau_{\text{subj}}(\tau_2) = \tau_{\text{subj}}(\tau_1)$$

$$\Delta\tau_{\text{subj}} = 0, \quad \Delta\tau_{\text{obj}} = \tau_2 - \tau_1 > 0$$

τ_{subj} — Subjective time reading assigned to consciously registered experience.

$\Delta\tau_{\text{subj}}$ — Subjectively experienced duration across the stipulated sleep interval.

$\Delta\tau_{\text{obj}}$ — Positive objective proper-time interval between sleep onset and awakening.

τ_1 — Objective proper-time reading at the beginning of the unexperienced interval.

τ_2 — Objective proper-time reading at the end of the unexperienced interval.

The zero subjective interval represents the instantaneous experienced transition from sleep onset to awakening, even though objective clocks advance throughout. It is a representation of the stated absence of registered experience, not a claim that all sleep stages lack experience. It introduces neither a second physical time dimension nor motion at light speed.

6. Conclusions

BUFT treats conventional superluminal travel as structurally prohibited. The local wave equation and assumed Lorentzian causal structure supply the propagation argument; the normalized velocity ansatz remains within the same domain. The finite rest-energy reference and temporal scaling are retained without assigning a physical singularity to the mass–energy equation.

Recursive field transformations and expansion geometry are distinguished from local transport. Consciousness and sleep concern the interpretation of experienced duration: an idealized zero subjective interval coexists with positive objective progression. Neither that experience nor a field-amplitude reweighting establishes a superluminal transport mechanism.

References

1. Einstein, A. (1905). Zur Elektrodynamik bewegter Körper. Annalen der Physik, 17, 891–921.

2. Einstein, A. (1916). Die Grundlage der allgemeinen Relativitätstheorie. Annalen der Physik, 49, 769–822.

3. Misner, C. W., Thorne, K. S., and Wheeler, J. A. (1973). Gravitation. W. H. Freeman.

4. Hawking, S. W., and Ellis, G. F. R. (1973). The Large Scale Structure of Space-Time. Cambridge University Press.

5. Carroll, S. M. (2004). Spacetime and Geometry: An Introduction to General Relativity. Addison-Wesley.

6. Penrose, R. (2004). The Road to Reality: A Complete Guide to the Laws of the Universe. Jonathan Cape.

7. Wheeler, J. A. (1990). A Journey into Gravity and Spacetime. Scientific American Library.

8. Tegmark, M. (1998). The interpretation of quantum mechanics: Many worlds or many words? Fortschritte der Physik, 46, 855–862.

—※—※—※—

Article 52

Equations of the Bakous Unified Field Theory: A Framework for Time-Space, Gravity, and Conscious Observation

Abstract

The following equations represent the mathematical framework of BUFT, integrating the principles of timespace curvature, gravity, electromagnetism, and conscious observation, as derived from the foundational $E = mc^2$.

1. Energy-Matter Equivalence in BUFT

$E = mc^2$

Purpose:

Describes the fundamental relationship between rest energy E and rest mass m, related by the vacuum light speed c. In BUFT, this equation serves as the constant and singularity governing timespace propagation and the Bakous Energy Field.

2. Time-Space Generation at T = 0

$$\Delta S(t) = \int_0^t c\,\mathrm{d}t' = ct$$

Purpose:
Defines spatial displacement ΔS accumulated over a finite elapsed time t by propagation at constant vacuum light speed c within the Bakous Energy Field. Here t′ is the integration variable.

3. Entropy, Gravity, and Time's Forward Motion

$$G \propto \frac{1}{T}$$

Purpose:
Indicates that gravity G is inversely proportional to temperature T, postulating that gravity drives time's forward movement and enables light propagation through timespace curvature.

4. Relationship Between Gravity and Magnetism

$$G \propto \frac{1}{B^2}$$

Purpose:
Relates gravitational field strength G to the inverse square of nonzero magnetic-field magnitude B, illustrating that gravity arises as an entropic effect of electromagnetism in BUFT.

5. Conservation of Energy and Conscious Observation

$$E_{obs} = \int \varphi \, dV$$

Purpose:

Represents the observed energy E_{obs} encoded in timespace by conscious observation, where φ is the electromagnetic flux and $\mathbf{dV}$ is the volume element.

This reflects the conservation of energy postulate, asserting that thoughts persist as encoded energy in the fabric of timespace.

6. Wave Properties of Time

$$t(x) = A \cdot sin(kx - \omega t)$$

Purpose:

Models time **t** as a wave-like phenomenon, where **A** is the amplitude, **k** is the wave number, ω is the angular frequency, and **x** is the spatial coordinate. This reflects the kinetic energy of time as observed in BUFT.

7. Consciousness as the Fifth Dimension

$$C = \frac{1}{\gamma \cdot E_{rad}}$$

Purpose:

Defines consciousness C as inversely proportional to the relativistic factor γ and the observed electromagnetic radiation energy E_{rad}, situating it within the fifth dimension in BUFT.

8. Energy Distribution of Harmonic Oscillating Strings

$$E = \frac{1}{2} \cdot k \cdot A^2$$

Purpose:

The energy **E** of harmonic oscillating strings that generate space is proportional to the amplitude **A** squared and the string constant **k**, driving the exponential expansion of the Bakous Energy Field.

9. The Expansion of Space with Each Eruption

$$V = V_0 \cdot e^{t/\tau}$$

Purpose:

The volume V grows from its initial value V_0 as elapsed time t increases. The positive characteristic time constant τ sets the exponential growth timescale for spatial expansion in BUFT.

10. Entropy and Time's Arrow

$$S = k_B \cdot ln(\Omega)$$

Purpose:

The entropy S of the universe, proportional to the natural logarithm of the number of microstates Ω, ensures the forward motion of time and the conservation of energy in BUFT.

11. Relativistic Time-Space Curvature and Mass Distribution

$$\varkappa = \frac{\Delta R}{M \cdot V}$$

Purpose:

Defines the curvature factor $\varkappa$ of timespace as a function of the change in curvature **ΔR**, mass **M**, and the expanding volume **V**. This equation links mass distribution and space generation in the Bakous Energy Field, emphasizing the role of curvature in propagating time forward.

12. Gravitational Time Dilation in BUFT

$$t' = t \cdot \sqrt{1 - \frac{2GM}{rc^2}}$$

Purpose:

Describes how gravitational fields influence time **t'** relative to the observer, based on gravitational constants **G**, mass **M**, and radial distance **r**. In BUFT, this equation shows that time dilation stems from the energy interactions within timespace curvature.

13. Electromagnetic Wave Dissipation in Time-Space

$$E_r = E_0 \cdot e^{(-\alpha r)}$$

Purpose:

Represents the dissipation of electromagnetic wave energy E_r as a function of distance **r** within timespace. The dissipation rate α depends on local curvature and gravitational effects, highlighting how electromagnetic energy interacts with timespace.

14. Oscillatory Expansion of Time-Space

$$S(t) = S_0 \cdot cos(\omega t) + \frac{1}{2} \cdot a \cdot t^2$$

Purpose:

Describes the oscillatory behavior of space S over time **t**, combining periodic oscillations **cos(ω t)** with a quadratic term that models accelerated expansion. This reflects harmonic string interactions in the Bakous Energy Field.

15. Quantum Vacuum Energy in BUFT

$$E_v = \frac{\hbar\omega}{2} \cdot \int_a^S \varrho_v \cdot dV$$

Purpose:

Defines vacuum energy $\mathbf{E_v}$ as a function of quantum harmonic oscillations $\hbar\omega$ and vacuum density ϱ_v. In BUFT, vacuum energy is directly tied to harmonic strings and their role in generating new volumes of timespace.

16. Conscious Observation and Quantum State Collapse

$$P_{obs} = |\psi|^2 / \int_{-\infty}^{+\infty} |\psi|^2 \, dx$$

Purpose:

Describes the probability $\mathbf{P_{obs}}$ of conscious observation collapsing a quantum state, where **ψ** is the wavefunction. In BUFT, this equation links the observer's consciousness to the dynamics of quantum systems within timespace.

17. Temporal Acceleration Due to Gravity

$$\ddot{t} = g \cdot \nabla R$$

Purpose:

Models the acceleration of time $\ddot{t}$ as a product of local gravitational strength g and the gradient of timespace curvature ∇R. This emphasizes gravity's role in propagating time forward within BUFT.

18. Conservation of Momentum in Time-Space Expansion

$$p = mv + m\frac{\partial \Delta S}{\partial t}$$

Purpose:

Models momentum p as mass m times the sum of velocity v and the spatial-expansion velocity $\partial\Delta S/\partial t$. Here ΔS is spatial displacement; both velocities are measured along the same chosen direction.

19. Interaction of Consciousness with Time-Space

$$C_{int} = \frac{E_{rad}}{k_B \cdot T}$$

Purpose:

Describes the interaction of consciousness C_{int} with timespace as a ratio of electromagnetic radiation energy E_{rad} to thermal energy $k_B \cdot T$. This highlights the thermodynamic connection between conscious observation and timespace dynamics.

20. Dynamic Relationship Between Energy and Time-Space Expansion

$$E_{\text{dyn}} = \alpha \cdot \frac{S}{t}$$

Purpose:

Models the dynamic relationship between energy E_{dyn} and timespace expansion, where α is a constant of proportionality, S is space, and t is time. This equation captures how expanding space dynamically generates energy, contributing to the continuous flow of time and the accelerating expansion of the Bakous Energy Field.

Summary

These equations provide a unified mathematical framework for the Bakous Unified Field Theory, revealing the profound connections between matter, energy, gravity, timespace, and consciousness. They challenge conventional paradigms and redefine the fundamental principles governing the universe.

—※—※—※—

Article 53

Testable Predictions of the Bakous Unified Field Theory (BUFT): A Framework for Empirical Validation

Abstract

The Bakous Unified Field Theory (BUFT) proposes a unified physical framework in which relativistic geometry, quantum behavior, and energetic propagation emerge through interactions within the Bakous Energy Field (BEF), a foundational energetic substrate underlying observable reality. This paper presents five proposed empirical signatures and observational protocols involving atomic-clock timing, gravitational lensing, cosmological polarization, neutrino oscillations, and ultra-high-energy cosmic rays.

Each test uses residual analysis: conventional relativistic and quantum contributions are modeled, with their uncertainties, before testing for a specified BEF-linked contribution. The resulting framework constrains parameterized BUFT relations through precision observations. Detectability depends on the predicted signal, instrumental sensitivity, and control of systematic effects.

1. Introduction

Modern physics remains divided between two extraordinarily successful frameworks: general relativity and quantum mechanics. General relativity accurately describes gravitational geometry and cosmological structure across large scales, while quantum mechanics predicts microscopic interactions with exceptional precision. Their unification remains an open problem in regimes requiring both strong gravity and quantum dynamics, including black-hole interiors and the earliest cosmological epochs.

The Bakous Unified Field Theory (BUFT) proposes that observable reality emerges through interactions between timespace geometry, energetic propagation, and the Bakous Energy Field (BEF), a foundational energetic substrate underlying relativistic and quantum phenomena alike. Within this framework, gravity is interpreted as timespace curvature generated through energetic gradients within the BEF itself.

Rather than attempting to discard established physics, BUFT treats general relativity and quantum mechanics as highly accurate emergent approximations arising from deeper energetic structure. Consequently, the central empirical objective of BUFT is not necessarily to overturn conventional theory, but to determine whether measurable observational signatures persist beyond standard relativistic and quantum predictions.

Each forecast is constructed around measurable quantities capable of direct empirical comparison against conventional models. The relations below specify parameterized departures; a quantitative test additionally requires the relevant BEF contribution to be fixed independently of the residual being tested.

2. Foundational BUFT Residual Structure

Within the BUFT framework, observable physical quantities may be represented as conventional relativistic or quantum predictions modified by a small BEF-dependent contribution:

$$O_{\text{obs}} = O_{\text{std}} + \delta_{\text{BEF}}$$

O_{obs} — Observed physical quantity derived from measurement or reconstructed data.

O_{std} — Prediction from the specified relativistic or quantum model.

δ_{BEF} — Proposed BEF contribution to the same observable.

All three terms refer to the same physical quantity and normalization. The empirical objective is to detect or constrain the specified BEF contribution through precision observation.

Each test uses a defined BEF template, coupling parameters, and an uncertainty model. Model comparisons assess residual origins. Null results constrain the tested coupling and template over
the experiment's sensitivity range; reproducible departures are checked against conventional explanations.

3. Prediction: BEF-Modulated Relativistic Time Dilation

Framework

General relativity predicts elapsed proper time from the metric and clock trajectories. BUFT proposes an additional contribution associated with local BEF-gradient structure.

MATHEMATICAL EXPRESSION

$$\Delta t_{\text{obs}} = \Delta t_{\text{GR}}(1 + \beta_{\text{BEF}}\Phi_{\text{BEF}})$$

Δt_{obs} – Measured elapsed proper-time interval for the clock comparison.

Δt_{GR} – Elapsed proper-time interval predicted for the same clock trajectory by general relativity.

β_{BEF} – Coupling governing the proposed BEF correction to temporal propagation.

Φ_{BEF} – Normalized effective BEF potential contribution for the measurement environment.

The product of the coupling and normalized potential is small in the perturbative regime, and the full correction factor remains positive. Gravity governs forward temporal progression throughout the comparison.

EXPERIMENTAL PROTOCOL

Two synchronized optical lattice atomic clocks are operated continuously within differing gravitational environments while environmental and relativistic variables are independently monitored and corrected throughout the observation interval. Continuous timing measurements are collected over extended durations to characterize stochastic fluctuations and isolate persistent timing structure.

The standard relativistic gravitational redshift prediction is first calculated using established general-relativistic methods for the clock trajectories. Experimental corrections are then applied for atmospheric variation, thermal drift, terrestrial motion, tidal deformation, instrumental instability, and synchronization uncertainty. The experimentally observed fractional frequency shift is subsequently compared against the predicted relativistic shift.

The measured fractional frequency shift is compared with the relativistic prediction, with statistical and systematic uncertainties propagated into the residual. Define y = Δf/f using the same frequency convention for both values. The relative timing residual is:

$$R_t = \frac{y_{obs} - y_{GR}}{y_{GR}}, \quad y_{GR} \neq 0$$

R_t – Relative deviation of the observed fractional frequency shift from the relativistic prediction.

y – Fractional frequency shift Δf/f for the clock comparison.

Δf – Frequency difference between the compared clocks.

f – Reference clock frequency used for normalization.

y_{obs} – Observed fractional frequency shift.

y_{GR} – Fractional frequency shift predicted by general relativity.

When the predicted shift vanishes, use the difference between observed and predicted fractional frequency shifts without dividing by the reference shift.

EXPECTED OBSERVATIONAL SIGNATURE

A nonzero BEF correction produces a systematic timing departure after conventional contributions are modeled. A stationary correction produces a rate offset; a time-dependent contribution can produce drift. The BEF model supplies the temporal profile used with the elapsed-time relation.

Repeated runs test whether the residual follows the independently specified BEF contribution between the clock environments. Coherence and reproducibility distinguish a candidate signal from stochastic noise, with systematic effects assessed separately.

4. Prediction: BEF-Induced Gravitational Lensing Residuals

FRAMEWORK

General relativity predicts gravitational lensing through spacetime curvature associated with matter-energy distributions. BUFT proposes an additional BEF curvature contribution capable of modifying photon trajectories.

MATHEMATICAL EXPRESSION

$$\alpha_{obs} = \alpha_{GR}(1 + \gamma_{BEF}\mathscr{T}^{\gamma}_{BEF})$$

α_{obs} – Observed lensing angle for the selected deflection component.

α_{GR} – Lensing angle predicted from the relativistic matter-distribution model.

γ_{BEF} – Coupling governing the BEF correction to lensing geometry.

$\mathscr{T}^{\gamma}_{BEF}$ – Normalized BEF path contribution along the photon trajectory.

The multiplicative correction recovers the relativistic angle when the BEF coupling vanishes. The same deflection convention is used for both angles.

EXPERIMENTAL PROTOCOL

A statistically suitable sample of galaxy-cluster lensing systems is selected from high-resolution weak-lensing survey datasets. Each cluster is reconstructed using the same observational and geometric conventions.

Each cluster is independently reconstructed using standard gravitational models incorporating baryonic matter distributions, dark-matter halo reconstruction, and conventional relativistic lensing geometry derived from observational shear mapping. The predicted relativistic lensing angle for each system is compared directly against the observed angle reconstructed from weak-lensing distortions, propagating reconstruction uncertainties.

Instrumental uncertainty, foreground contamination, reconstruction bias, line-of-sight structure, and observational noise are modeled in the comparison. The remaining lensing deviation is defined as:

$$R_\alpha = \frac{\alpha_{\text{obs}} - \alpha_{\text{GR}}}{\alpha_{\text{GR}}}, \qquad \alpha_{\text{GR}} \neq 0$$

R_α — Fractional deviation between observed and predicted lensing angles.

For a vanishing reference angle, use the angular difference directly. Within the stated model, the fractional residual equals the product of the BEF coupling and normalized photon-path contribution.

EXPECTED OBSERVATIONAL SIGNATURE

If BEF structure contributes directly to curvature behavior, reconstructed lensing geometries exhibit departures from the conventional matter-distribution model. Cluster shear maps can test for coherent angular amplification or directional curvature offsets associated with the specified BEF path contribution.

A preference for dynamically complex environments or strong energetic-density gradients is a proposed pattern to test. The chosen BEF path model determines the angular form of the shear prediction.

5. Prediction: BEF-Induced Electromagnetic Polarization Rotation

Framework

BUFT proposes that electromagnetic waves propagating through varying BEF-density regions experience polarization rotation beyond conventional plasma and magnetic-field effects. The proposed contribution is tested after known electromagnetic effects are modeled.

Mathematical Expression

$$\theta_{\mathrm{obs}} = \theta_{\mathrm{EM}} + \theta_{\mathrm{BEF}}$$

$$\theta_{\mathrm{BEF}} = \kappa_{\mathrm{BEF}} \int_{\mathscr{C}} \nabla \rho_{\mathrm{BEF}} \cdot d\ell$$

θ_{obs} – Observed electromagnetic polarization rotation angle.

θ_{EM} – Rotation predicted from conventional electromagnetic propagation effects.

θ_{BEF} – Proposed polarization rotation associated with BEF-density variation.

κ_{BEF} – Coupling converting the integrated normalized density gradient into polarization rotation.

ρ_{BEF} – Normalized local BEF-density field sampled along the propagation path.

$\mathscr{C}$ – Oriented electromagnetic propagation path from source to detector.

$d\ell$ – Directed spatial displacement along the propagation path.

For a differentiable, time-independent density field and constant coupling, the gradient integral is the density difference between the path endpoints:

$$\theta_{BEF} = \kappa_{BEF}[\rho_{BEF}(B) - \rho_{BEF}(A)]$$

A — Source endpoint of the electromagnetic path.

B — Detector endpoint of the electromagnetic path.

Under these conditions, intermediate positive and negative gradient contributions cancel. Equal endpoint densities give zero rotation in this expression. Time-dependent propagation requires the density field along the actual path to be specified explicitly.

Experimental Protocol

Cosmological polarization datasets are analyzed after independently modeling known contributions from galactic dust, synchrotron emission, plasma Faraday rotation, detector polarization offsets, and instrumental calibration drift. Multiple foreground-removal algorithms are applied independently to assess model-dependent contamination within the final polarization maps.

The polarization rotation angle is reconstructed through comparison between observed polarization orientation and the polarization state predicted through a specified conventional propagation and source-polarization model. Independent sky regions are analyzed separately to determine whether directional rotational structure remains spatially coherent across cosmological scales.

Statistical and systematic uncertainties are retained in the residual analysis. Foreground modeling and calibration tests establish whether an apparent rotation survives independent processing procedures.

The remaining polarization deviation is defined as:

$$R_\theta = \theta_{obs} - \theta_{EM} = \theta_{BEF}$$

R_θ — Polarization rotation remaining after the conventional electromagnetic contribution is subtracted.

EXPECTED OBSERVATIONAL SIGNATURE

The proposed observable is a residual rotation correlated with the specified BEF-density field. For the time-independent gradient model, the correlation is with the source-to-detector endpoint density difference. This integral describes the stated endpoint contribution; intermediate-gradient rotation is evaluated through the corresponding path model. Independent sky regions and foreground-removal procedures test for reproducible angular offsets. Parity asymmetries, directional patterns, and frequency dependence must be calculated from the specified den- sity and propagation model before they are used as distinguishing signatures. Plasma and magnetic-field structure are included in the comparison.

6. Prediction: BEF-Modulated Neutrino Oscillation Residuals

FRAMEWORK

Quantum mechanics predicts neutrino oscillation probabilities through flavor mixing and mass splitting. BUFT proposes that BEF structure weakly modulates these probabilities during neutrino propagation.

MATHEMATICAL EXPRESSION

$$P_{\text{obs}} = P_{\text{QM}} + \varepsilon_{\text{BEF}}\mathcal{T}^{\nu}_{\text{BEF}}$$

Both probabilities refer to the same initial and detected flavors, neutrino energy, baseline, and propagation environment.

P_{obs} — Observed flavor-transition probability inferred from detector events.

P_{QM} — Transition probability predicted by the standard quantum oscillation model.

ν_α — Initial neutrino flavor state.

ν_β — Detected neutrino flavor state.

ε_{BEF} — Coupling governing the proposed BEF correction to neutrino propagation.

$\mathcal{T}^{\nu}_{\text{BEF}}$ — Normalized BEF path contribution along the neutrino trajectory.

The correction is restricted so that each transition probability remains between zero and one. Across all detected flavors, the corrected probabilities sum to one. These bounds constrain the allowed channel corrections.

EXPERIMENTAL PROTOCOL

Atmospheric neutrino datasets are analyzed across multiple propagation baselines. Standard mixing angles, mass splittings, matter effects, and detector responses are included in a consistent fit, retaining correlations with the proposed BEF correction.

Event reconstruction algorithms are used to determine flavor-transition probabilities across varying propagation distances and traversal geometries. The predicted oscillation probabilities derived from standard quantum oscillation theory are compared directly against experimentally reconstructed transition probabilities. Statistical and systematic uncertainties from detector response, atmospheric flux variation, background contamination, and reconstruction bias are propagated into the deviation analysis.

The flavor-transition residual is:

$$R_\nu = P_{\text{obs}} - P_{\text{QM}}$$

R_ν — Flavor-transition probability residual relative to the standard oscillation prediction.

EXPECTED OBSERVATIONAL SIGNATURE

If BEF structure contributes to neutrino propagation, the residual follows the specified neutrino-path contribution across energy and baseline. Stable transition-amplitude distortions or propagation-dependent asymmetries are candidate signatures to test against conventional oscillation uncertainties.

Phase displacement is modeled through an energy- and baseline-dependent correction, distinguished from an additive probability contribution. The same BEF model must reproduce any
claimed directional consistency across independent detector events and datasets.

7. Prediction: BEF-Induced Cosmic-Ray Energy Spectrum Residuals

FRAMEWORK

BUFT proposes that ultra-high-energy cosmic rays traversing varying BEF-density regions experience energetic perturbations during propagation.

Mathematical Expression

$$E_{\mathrm{obs}} = E_0(1 + \eta_{\mathrm{BEF}}\mathcal{T}_{\mathrm{BEF}}^{\mathrm{CR}})$$

E_{obs} — Cosmic-ray energy reconstructed from detector-array event analysis.

E_0 — Expected energy from the standard source and propagation model.

η_{BEF} — Coupling governing the BEF correction to cosmic-ray energy propagation.

$\mathcal{T}_{\mathrm{BEF}}^{\mathrm{CR}}$ — Normalized BEF path contribution along the cosmic-ray trajectory.

The expected energy is positive, and the multiplicative factor preserves positive reconstructed energy. Vanishing BEF coupling recovers the standard prediction.

Experimental Protocol

Ultra-high-energy cosmic-ray events are grouped by reconstructed arrival direction and inferred propagation trajectory. Magnetic deflection, detector acceptance bias, atmospheric interaction effects, source-distribution uncertainty, and composition are modeled in the analysis. Event populations are then statistically categorized according to reconstructed large-scale propagation environments.

The experimentally reconstructed cosmic-ray energies are compared statistically against predictions from conventional propagation models incorporating known interaction mechanisms and astrophysical source distributions. Source energies and trajectories enter through those models and their uncertainties; they are not independently known for individual events.

The fractional energetic deviation is defined relative to the positive standard-model energy:

$$R_E = \frac{E_{obs} - E_0}{E_0}$$

R_E — Fractional deviation between reconstructed and conventionally predicted cosmic-ray energy.

EXPECTED OBSERVATIONAL SIGNATURE

If ultra-high-energy propagation is influenced by BEF-density structure, reconstructed spectra exhibit departures correlated with the specified propagation contribution. Directional spectral distortion and stable arrival-energy offsets are candidate patterns to test across independent event populations.

The sign of an energy shift is determined by the BEF coupling and path contribution. The fitted coupling determines whether the predicted response is amplification or suppression. Source populations, composition, magnetic deflection, and detector exposure must be included when assessing a residual anisotropy.

8. Conclusion

The Bakous Unified Field Theory proposes that observable relativistic and quantum phenomena emerge from deeper energetic structure associated with the Bakous Energy Field. The five tests presented here formulate parameterized deviations for observational comparison with established physics.

Each test combines measurable quantities, a specified BEF contribution, and an uncertainty model. Null results constrain the tested parameters at the achieved sensitivity. Reproducible departures across independent observations can support further investigation when the same model accounts for them and conventional explanations have been assessed.

The empirical validity of BUFT rests on quantitative agreement with observation. The protocols organize detection and validation through comparisons among predicted templates, measured residuals, and experimental controls.

References

1. Bakous, R. (2025). Bakous Unified Field Theory: A Timespace Perspective. Independently Published.

2. Einstein, A. (1916). Die Grundlage der allgemeinen Relativitätstheorie. Annalen der Physik 49, 769–822.

3. Planck Collaboration. (2020). Planck 2018 results. VI. Cosmological parameters. Astronomy & Astrophysics 641, A6.

4. Madhavacheril, M. S. et al. (2024). The Atacama Cosmology Telescope: DR6 Gravitational Lensing Map and Cosmological Parameters. The Astrophysical Journal 962, 113.

5. Euclid Collaboration. (2020). Euclid preparation. VII. Forecast validation for Euclid cosmological probes. Astronomy & Astrophysics 642, A191.

6. IceCube Collaboration. (2018). Measurement of Atmospheric Neutrino Oscillations at 6–56 GeV with IceCube DeepCore. Physical Review Letters 120, 071801.

7. Pierre Auger Collaboration. (2020). Features of the Energy Spectrum of Cosmic Rays above 2.5×10^{18} eV Using the Pierre Auger Observatory. Physical Review Letters 125, 121106.

8. Bothwell, T. et al. (2022). Resolving the gravitational redshift across a millimetre-scale atomic sample. Nature 602, 420–424.

9. Hyper-Kamiokande Collaboration. (2018). Hyper-Kamiokande Design Report. arXiv:1805.04163.

—※—※—※—

ARTICLE 54

Quantized Harmonic Oscillations and the Emergence of the Yang-Mills Mass Gap: A Solution via the Bakous Unified Field Theory (BUFT)

ABSTRACT

The Bakous Unified Field Theory (BUFT) presents a unified geometric resolution to the Yang-Mills mass-gap problem through harmonic confinement within the Bakous Energy Field (BEF). In BUFT, non-Abelian field excitations emerge as quantized harmonic modes embedded within an eight-dimensional timespace manifold governed by the BEF. This manifold consists of three spatial dimensions, time, consciousness, entropy, the Bakous Energy Field, and the governing laws of physics unified within a single geometric structure. The discrete organization of these oscillatory states generates a nonzero vacuum excitation threshold, producing positive spectral separation while preserving gauge symmetry.

The framework restricts physical excitations to discrete energetically stable configurations determined by the geometric and harmonic structure of timespace itself. Through intrinsic harmonic confinement, the excitation spectrum acquires a natural positive lower bound, yielding confinement and mass-gap emergence directly from the energetic resistance and quantized curvature structure of the BEF. In this way, BUFT unifies gauge stability, spectral discretization, and nonzero vacuum excitation thresholds within a single geometric framework.

1. Introduction

The Yang-Mills mass-gap problem concerns the existence of a quantum Yang-Mills theory with compact simple gauge group **G** on $\mathbb{R}^4$ possessing a strictly positive lower bound between the vacuum state and the first nontrivial excitation state.

Formally, the theory requires:

$$\Delta > 0$$

where:

Δ — minimum non-vacuum excitation energy (mass gap)

Conventional approaches rely upon perturbative quantum field theory, lattice gauge methods, and nonperturbative confinement analysis. Within BUFT, however, the problem is reformulated geometrically through the structure of timespace itself.

BUFT models reality as an eight-dimensional timespace manifold unifying three spatial dimensions, time, consciousness, entropy, the Bakous Energy Field, and the governing laws of physics within a single coherent geometric architecture. Within this framework, the Bakous Energy Field functions as the foundational energetic substrate governing field interactions, harmonic structure, and geometric evolution throughout timespace.

Rather than treating fields as isolated primitives, BUFT interprets non-Abelian gauge excitations as quantized harmonic oscillatory states emerging from the Bakous Energy Field. Harmonic confinement within the BEF restricts field excitations to discrete stable modes, naturally generating nonzero energetic separation across the excitation spectrum.

Under this framework, confinement is not imposed externally through symmetry violation or arbitrary boundary conditions. Instead, it emerges intrinsically from the energetic resistance, curvature structure, and harmonic constraints of the BEF.

2. Formal Yang-Mills Structure

Consider a compact simple gauge group G with gauge connection:

$$A_\mu^a(x)$$

where:

$A_\mu^a(x)$ — Yang-Mills gauge connection field

$\mu \in \{0, 1,2,3\}$ — Coordinate index for the four-dimensional field components.

a — Lie algebra generator index associated with gauge group G

The Yang-Mills field-strength tensor is defined by:

$$F_{\mu\nu}^a = \partial_\mu A_\nu^a - \partial_\nu A_\mu^a + g f^{abc} A_\mu^b A_\nu^c$$

where:

$F_{\mu\nu}^a$ — non-Abelian Yang-Mills field-strength tensor

∂_μ — relativistic field-space derivative operator

g — gauge coupling constant

f^{abc} — structure constants of the gauge algebra

The effective four-dimensional Yang-Mills projection embedded within BUFT is represented by the action:

$$S_{YM} = -\frac{1}{4}\int F^a_{\mu\nu}F^{a\mu\nu}\, d^4x$$

where:

S_{YM} — Yang-Mills action functional

d^4x — effective four-dimensional relativistic field-space volume element embedded within the larger eight-dimensional timespace manifold

Within BUFT, the conventional four-dimensional integration measure represents the observable projection of gauge-field dynamics onto relativistic field space arising from the deeper geometry of timespace itself.

The theory admits a Hilbert space:

$$\mathcal{H}$$

with Hamiltonian operator:

$$\hat{H}$$

acting upon physical excitation states.

$\mathcal{H}$ — Hilbert space of physical field states

$\hat{H}$ — Hamiltonian operator governing system evolution

The mass-gap condition requires:

$$spec(\hat{H}) \subset \{0\} \cup [\Delta,\infty)$$

with $\Delta > 0$

where:

$spec(\hat{H})$ — spectrum of allowable Hamiltonian energy eigenvalues

$\{0\}$ — vacuum energy state

$[\Delta,\infty)$ — positive excitation spectrum above the vacuum

Thus all non-vacuum excitations possess strictly positive minimum energy.

3. Harmonic Quantization in BUFT

Within BUFT, non-Abelian field excitations are interpreted as discrete harmonic oscillatory modes embedded within the Bakous Energy Field.

The excitation spectrum is governed by:

$$E_n = n\hbar\omega + \Delta\,, \quad (n \geq 1)$$

where:

E_n — excitation energy of harmonic mode n

$n \in \mathbb{N}$ — discrete excitation index above the vacuum state

$\hbar$ — reduced Planck constant

ω — Angular frequency of the harmonic mode.

Δ — Positive additive energy offset in the excitation model.

Dimensional consistency follows directly since:

$$[\hbar\omega] = [\Delta] = [energy]$$

where:

[ ] — Physical dimension of the indicated quantity.

The vacuum state is defined by:

$E_{vac} = 0$

where:

E_{vac} — vacuum-state energy

This yields a discrete excitation spectrum bounded below by a positive energetic threshold above the vacuum state.

Within the BUFT framework, this threshold emerges from intrinsic resistance of the BEF to stable harmonic excitation. Only oscillatory states satisfying the geometric and energetic constraints of timespace remain dynamically stable across the excitation spectrum.

4. Harmonic Confinement and Spectral Separation

BUFT interprets confinement as a direct consequence of harmonic restriction within the BEF.

The Bakous Energy Field functions as a structured energetic substrate whose curvature and harmonic geometry permit only discrete stable excitation states. Continuous unconstrained field configurations become energetically suppressed, while geometrically stable harmonic modes persist as quantized excitations.

Under this structure:

- *non-Abelian excitations cannot occupy arbitrarily small positive energies*
- *vacuum transitions require finite excitation energy*

- *spectral continuity near the vacuum state becomes prohibited*

Nonzero energetic separation therefore emerges naturally:

$$E_n - E_{vac} = n\hbar\omega + \Delta$$

for all nontrivial excitation states:

$$n \geq 1$$

where

$E_n - E_{vac}$ — energetic separation above the vacuum state

The excitation spectrum remains discretized above the vacuum configuration, producing harmonic confinement and a nonzero Yang-Mills mass gap through the energetic geometry of timespace itself.

5. Gauge Symmetry Embedding in BUFT

BUFT preserves gauge symmetry through harmonic correspondence between excitation states and stable representations of the gauge algebra.

Let:

$$\psi_n^a$$

denote harmonic excitation states associated with generator index a.

where:

ψ_n^a — harmonic excitation state associated with gauge generator a

The excitation structure preserves algebraic closure through the Lie algebra commutation relations:

$$[T^a, T^b] = if^{abc}T^c$$

where:

T^a	–	generators of the gauge algebra
$[T^a, T^b]$	–	commutator operation between generators
i	–	imaginary unit satisfying $i^2 = -1$

Within BUFT, harmonic confinement restricts the physically realizable excitation spectrum while preserving gauge covariance and algebraic consistency across all stable excitation modes.

The harmonic structure therefore operates not as a symmetry-breaking mechanism, but as a spectral-selection mechanism imposed directly by the energetic curvature and geometric constraints of the BEF.

6. Operator Structure and Spectral Stability

Within the BUFT framework, the Hamiltonian operator governing harmonic field excitations is represented schematically by:

$$\hat{H}_{BUFT} = \sum_n (n\hbar\omega + \Delta)\,|n\rangle\,\langle n|$$

where:

$\hat{H}_{BUFT}$	–	BUFT harmonic Hamiltonian operator
$\sum_n$	–	summation across all allowable excitation states
$|n\rangle\,\langle n|$	–	projection operator onto excitation state n
$|n\rangle$	–	quantized harmonic excitation state
$\Delta > 0$	–	positive vacuum separation threshold

The resulting excitation spectrum satisfies:

$$spec(\hat{H}_{BUFT}) \subset \{0\} \cup [\Delta, \infty)$$

for all non-vacuum states.

The positivity of the spectrum emerges from harmonic confinement within the BEF itself. Because physically realizable field configurations are restricted to energetically stable oscillatory states, unconstrained vacuum collapse and continuous low-energy excitation modes remain prohibited.

This produces stable energetic discreteness across the non-Abelian excitation structure while preserving gauge consistency throughout the harmonic field system.

7. Conclusion

The Bakous Unified Field Theory provides a unified geometric resolution to the Yang-Mills mass-gap problem through harmonic confinement within the Bakous Energy Field. Within this framework, non-Abelian field excitations emerge as discrete harmonic oscillatory states whose stability is governed by the energetic geometry and curvature structure of timespace itself. Harmonic confinement within the BEF restricts physically realizable field configurations to energetically stable excitation modes, naturally generating positive spectral separation above the vacuum state while preserving gauge symmetry across the excitation structure.

The resulting harmonic spectrum produces nonzero excitation thresholds directly from the intrinsic resistance and quantized structure of the BEF rather than from externally imposed boundary conditions or symmetry-breaking mechanisms. By embedding gauge structure, harmonic quantization, confinement, and spectral stability within a unified timespace geometry, BUFT establishes a coherent framework through which Yang-Mills spectral separation emerges naturally from the energetic architecture of the field itself.

References

1. Bakous, R. (2025). *Bakous Unified Field Theory: A Timespace Perspective.* Independently Published.
2. Jaffe, A., & Witten, E. (2026). *Quantum Yang-Mills Theory.* Clay Mathematics Institute Millennium Prize Problems.
3. Yang, C. N., & Mills, R. L. (1954). *Conservation of Isotopic Spin and Isotopic Gauge Invariance.* Physical Review, 96(1), 191-195.
4. Planck, M. (1900). *On the Theory of the Energy Distribution Law of the Normal Spectrum.*
5. Einstein, A. (1905). *Does the Inertia of a Body Depend Upon Its Energy Content?*
6. Wheeler, J. A. (1962). *Geometrodynamics.*
7. Peskin, M. E., & Schroeder, D. V. (1995). *An Introduction to Quantum Field Theory.*
8. Weinberg, S. (1995). *The Quantum Theory of Fields,* Vol. I.

—※—※—※—

ARTICLE 55

BUFT Interpretation of the Riemann Hypothesis: Prime Number Distribution in BUFT

ABSTRACT

This paper presents a BUFT interpretation of the Riemann Hypothesis within a physical framework describing fundamental interactions through an eight-dimensional timespace manifold governed by the Bakous Energy Field (BEF). The critical-line condition Re(s) = ½ is associated with a null-resonance postulate involving consciousness curvature and gravitational time-force modulation.

A field-response convolution supplies the mathematical resonance model. Under the stated regularity and symmetry assumptions, its normalized response is odd and vanishes at the centered origin. This result establishes a cancellation condition within the model. The proposed association with the Riemann zeta function remains a BUFT postulate, distinct from a proof locating all of its nontrivial zeros.

1. Introduction

The Riemann Hypothesis posits that all nontrivial zeros of the Riemann zeta function lie on the line Re(s) = ½. Traditionally examined in pure mathematics, this paper adopts a perspective from theoretical physics, specifically within the Bakous Unified Field Theory.

Under BUFT, the spectral properties of ζ(s) are interpreted through physical field interference within the higher-dimensional timespace structure. The zeta function retains its standard mathematical definition; the physical resonance construction below is a separate model whose proposed association with ζ(s) is a BUFT postulate.

2. Theoretical Framework: BUFT Overview

BUFT postulates that physical reality unfolds within an eight-dimensional timespace continuum. Its first four coordinates describe three spatial directions and one time dimension. The remaining dimensions represent consciousness curvature, entropy, the Bakous Energy Field, and the laws of physics:

Ψ_5 – Consciousness-curvature field associated with the fifth dimension.

S_6 – Entropy represented as directional tension in the sixth dimension.

$\mathscr{B}$ – Bakous Energy Field associated with the seventh dimension.

$\mathscr{L}_8$ – Representation of the laws of physics in the eighth dimension.

The interaction of forward-propagating time-force with consciousness curvature gives rise to quantized harmonic interference patterns, proposed within BUFT to underlie prime-number distribution and the spectral properties of ζ(s).

3. BUFT Resonance Convolution

For fixed curvature, entropy, and BEF conditions, define the field response by the convolution:

$$A(t) = \int_{-\infty}^{\infty} \Psi_5(u)\, R(\mathscr{B}, t - u)\, du$$

A – Physical field-response amplitude at the response coordinate t.

Ψ_5 – Consciousness-curvature field at the fixed curvature and entropy parameters.

R – BEF response kernel as a function of temporal displacement.

$\mathscr{B}$ – Fixed BEF state entering the response kernel.

t – Temporal coordinate of the modeled response.

u – Temporal integration variable for the curvature field.

This expression represents the field resonance feedback proposed in BUFT: the BEF reflects and superimposes patterns originating from consciousness-encoded curvature. Suppress the fixed BEF argument in subsequent formulas.

$$F(t) = \frac{A(t)}{A_{\mathrm{ref}}}, \quad A_{\mathrm{ref}} > 0$$

F – Normalized field-response amplitude.

A_{ref} – Fixed positive reference amplitude for the physical field response.

The normalized response is kept distinct from $\zeta(s)$. Its temporal coordinate has no prescribed identification with the complex zeta argument.

4. BUFT Interpretation of the Critical Line

Within BUFT, the critical-line condition is associated with balanced, opposite-phase field contributions. This association is a null-resonance postulate within the eight-dimensional manifold:

$$\mathrm{Re}(s) = \frac{1}{2}$$

$$s = \sigma + i\nu$$

s — Complex argument of the Riemann zeta function.

σ — Real part of the complex zeta argument.

ν — Imaginary part of the complex zeta argument.

i — Imaginary unit satisfying $i^2 = -1$.

The opposing contributions represent phase reflection in the resonance model. Negative field arguments express mathematical reflection about the centered origin; they do not describe backward physical temporal progression.

The convolution symmetry proved below establishes cancellation at that origin. The critical-line association remains the BUFT postulate; identifying resonance zeros with zeta zeros requires a specified mathematical correspondence preserving their locations.

5. Mathematical Properties of the Response

5.1. Function Domains

Assume that the consciousness-curvature field is smooth, bounded, and square-integrable, while the response kernel is absolutely integrable and decays exponentially. These assumptions define the convolution independently of its proposed physical interpretation.

With the fixed field parameters suppressed, the assumptions are:

$$\Psi_5 \in \mathrm{L}^2(\mathbb{R}), \quad R \in \mathrm{L}^1(\mathbb{R})$$

$$|R(\tau)| \leq C_\mathrm{R} \exp(-\lambda|\tau|), \quad \lambda > 0$$

τ – Temporal displacement in the response kernel.

C_R – Positive amplitude bounding the response kernel.

λ – Positive decay rate of the kernel envelope.

Boundedness of the field and absolute integrability of the kernel ensure absolute convergence at every response coordinate. Young's convolution inequality also places the response in the square-integrable class:

$$\|A\|_2 \leq \|\Psi_5\|_2 \|R\|_1$$

5.2. Symmetry and Cancellation

Suppose the field is odd and the kernel is even:

$$\Psi_5(-u) = -\Psi_5(u), \quad R(-\tau) = R(\tau)$$

Changing the integration variable from u to −u then gives:

$$A(-t) = -A(t)$$

$$F(-t) = -F(t) \quad \Rightarrow \quad F(0) = 0$$

This is a symmetry-enforced zero of the normalized response at the centered origin. Oddness does not make the convolution vanish at every coordinate or establish the locations of zeta zeros.

5.3. Normalization

$$[A_\mathrm{ref}] = [A] = [\Psi_5][R][t]$$

The reference amplitude matches the convolution's physical dimensions. The normalized response F remains distinct from ζ(s), which retains its mathematical normalization.

6. Prime Number Distribution in BUFT

Under BUFT, prime numbers are interpreted as discrete echoes of BEF field resonance. The proposed association relates the spectral zeros of $\zeta(s)$ to null points in field-curvature feedback. The cancellation result supplies a precise symmetry property of the resonance model; the connection to prime-number distribution remains the stated physical interpretation.

7. Conclusion

Through the lens of BUFT, the Riemann Hypothesis is interpreted as a critical-line null-resonance condition in timespace geometry, embedded in the laws of physics represented by the eighth dimension. Forward temporal progression is preserved throughout this account.

The proved mathematical result is the odd symmetry and centered cancellation of the normalized convolution under the specified assumptions. The BUFT critical-line association is retained as a postulate. Establishing the Riemann Hypothesis requires a correspondence with the standard zeta function that accounts for all nontrivial zeros.

References

1. Riemann, B. (1859). Ueber die Anzahl der Primzahlen unter einer gegebenen Grösse. Monatsberichte der Berliner Akademie.

2. Titchmarsh, E. C. (1986). The Theory of the Riemann Zeta-Function. Second edition, revised by D. R. Heath-Brown. Oxford University Press.

3. Edwards, H. M. (2001). Riemann's Zeta Function. Dover Publications.

4. Bakous, R. (2025). Bakous Unified Field Theory: A Timespace Perspective. Independently Published.

5. Connes, A. (1999). Trace formula in noncommutative geometry and the zeros of the Riemann zeta function. Selecta Mathematica (N.S.) 5, 29–106.

6. Berry, M. (1989). Quantum chaology, not quantum chaos. Physica Scripta 40, 335–336.

—※—※—※—

ARTICLE 56

BUFT Harmonic-Timespace Formalism for Four Millennium Prize Problems

ABSTRACT

This paper presents a unified BUFT-based resolution framework for four Millennium Prize Problems: The Birch and Swinnerton-Dyer conjecture, the Hodge conjecture, Navier–Stokes existence and smoothness, and P versus NP. Within the Bakous Unified Field Theory, these problems are interpreted not as isolated mathematical anomalies, but as lower-dimensional manifestations of deeper harmonic-timespace dynamics governed by entropy-curvature evolution, BEF-sourced stabilization, and BOFA tensor regulation. The framework establishes that elliptic rank behavior, algebraic projection stability, nonlinear vorticity suppression, and computational accessibility emerge from common geometric constraints embedded within the eight-dimensional timespace manifold. By introducing normalized BUFT field variables, BOFA tensor dynamics, harmonic resonance operators, projection functionals, and informational accessibility measures, the paper develops a generalized geometric framework through which these systems resolve within the broader harmonic-timespace structure of BUFT.

1. Introduction

The **Millennium Prize Problems** represent some of the deepest unresolved structures in mathematics and mathematical physics. Conventional approaches generally treat these problems through separate disciplines, including number theory, algebraic geometry, nonlinear analysis, and computational complexity. Within the **Bakous Unified Field Theory**, however, these divisions emerge as lower-dimensional consequences of incomplete harmonic-timespace description.

BUFT begins from a different foundational structure. Time is primary. Space emerges through temporal progression. Entropy governs structural ordering. Consciousness participates as a dimensional field. The Bakous Energy Field functions as the underlying energetic substrate from which timespace geometry emerges, while the laws of physics operate as the governing eighth-dimensional constraint structure.

From this perspective, the four problems addressed in this paper share a common origin: each contains an instability, divergence, incompleteness, or accessibility limitation that emerges when the system is analyzed outside the full BOFA-regulated manifold structure. The paper does not treat the Millennium Problems as metaphorical analogies, but as lower-dimensional manifestations of generalized harmonic-timespace accessibility dynamics.

This paper reformulates each problem within the BUFT framework and applies the BOFA tensor as the governing stabilization operator linking entropy-curvature evolution, BEF density, temporal progression, and emergent mathematical structure. The present treatment is developed as a generalized geometric formalism establishing structural equivalences and stabilization relations within the harmonic-timespace dynamics of BUFT.

2. Mathematical Foundations of the BUFT Manifold

Let $\mathscr{M}^8$ denote the ordered BUFT harmonic-timespace manifold:

$$\mathscr{M}^8 := \left(\mathscr{X}^1, \mathscr{X}^2, \mathscr{X}^3, \mathscr{T}, \Psi, \Sigma, \mathscr{E}_{BEF}, \Lambda\right)$$

where:

$\mathscr{X}^1$ *to* $\mathscr{X}^3$	–	represent the three emergent spatial dimensions
$\mathscr{T}$	–	represents the primary time dimension
Ψ	–	represents the consciousness dimension
Σ	–	represents entropy as a dimensional ordering field
$\mathscr{E}_{BEF}$	–	represents the Bakous Energy Field
Λ	–	represents the dimensional structure governing the laws of physics

The BUFT manifold is not a conventional spacetime extension. It is a harmonic-timespace hierarchy in which temporal progression precedes and generates emergent spatial structure through BEF-sourced entropy-curvature evolution.

To preserve harmonic scaling consistency throughout the generalized manifold, normalized BUFT fields are defined by:

$$\mathscr{E}\hat{} := \frac{\mathscr{E}_{BEF}}{\mathscr{E}_0}$$

$$\Sigma\hat{} := \frac{\Sigma}{\Sigma_0}$$

$$\mathscr{T}\hat{} := \frac{\mathscr{T}}{\mathscr{T}_0}$$

where:

$\mathscr{E}_0$ — denotes the BEF reference density scale

Σ_0 — denotes the entropy-curvature reference scale

$\mathscr{T}_0$ — denotes the temporal-force reference scale

The generalized BUFT action functional is:

$$\mathcal{S} := \int (\mathscr{R} + \mathscr{L}_{BEF} + \mathscr{L}_{\Sigma} + \mathscr{L}_{\Psi} + \mathscr{L}_{\Lambda}) d^8 \mathscr{X}$$

where:

$\mathscr{R}$ — represents generalized harmonic curvature

$\mathscr{L}_{BEF}$ — represents BEF field dynamics

$\mathscr{L}_{\Sigma}$ — represents entropy-curvature evolution

$\mathscr{L}_{\Psi}$ — represents consciousness-field structure

$\mathscr{L}_{\Lambda}$ represents law-governing dimensional constraints

The stabilizing admissibility conditions of the BUFT manifold are:

$$\int |\mathscr{E}\hat{}|^2 \, d\mathscr{V} < \infty$$

and:

$$|\nabla\Sigma\hat{}| < \mathscr{C}$$

for finite constant $\mathscr{C}$.

These conditions establish bounded BEF density and bounded entropy-curvature evolution throughout harmonic-timespace dynamics.

Observable lower-dimensional metric structure emerges as a stabilized projection of BOFA-regulated harmonic-timespace evolution rather than as a primary independent geometric background.

3. The BOFA Tensor

The BOFA tensor governs the coupling between BEF density, entropy-curvature evolution, temporal progression, and emergent lower-dimensional geometry throughout the generalized harmonic-timespace manifold.

Define the generalized BOFA tensor:

$$\mathcal{B}_{AB} := \kappa_B(\mathcal{P}_A \otimes \mathcal{P}_B) + \nabla_A \nabla_B \hat{\Sigma} + \lambda \mathcal{g}_{AB} |\mathcal{E}|^2$$

where:

$\mathcal{B}_{AB}$	–	denotes the BOFA tensor
κ_B	–	denotes BEF harmonic coupling strength
$\mathcal{P}_A$ and $\mathcal{P}_B$	–	are harmonic curvature vectors
$\otimes$	–	denotes tensor-product structure
$\nabla_A \nabla_B \hat{\Sigma}$	–	denotes entropy-curvature coupling
λ	–	denotes harmonic stabilization strength
$\mathcal{g}_{AB}$	–	denotes the emergent projected geometric coupling structure
$\mathcal{E}$	–	denotes normalized BEF density

The BOFA stabilization condition does not arise as an externally imposed constraint introduced to suppress instability artificially. Within BUFT, the positivity structure of the BOFA tensor follows from bounded entropy-curvature evolution together with finite normalized BEF-density scaling throughout admissible harmonic-timespace manifolds. The stabilization condition therefore emerges geometrically as a necessary admissibility relation for stable manifold evolution rather than as an externally imposed regularization principle.

For any admissible tangent vector $\mathscr{V}^A$:

$$\mathscr{V}^A \mathscr{B}_{AB} \mathscr{V}^B \geq 0$$

The positivity condition therefore expresses the requirement that physically admissible harmonic-timespace trajectories remain bounded under BOFA-regulated entropy-curvature evolution.

Under this formulation, observable lower-dimensional geometry does not evolve independently of BOFA stabilization dynamics. Metric structure emerges as a stabilized projection of generalized harmonic-timespace evolution governed through BEF density coupling and entropy-curvature regulation.

The BOFA tensor consequently governs:

- harmonic stabilization
- nonlinear accessibility structure
- resonance equilibrium
- informational-curvature evolution
- projection stability
- emergent lower-dimensional manifold geometry

4. Birch and Swinnerton-Dyer Conjecture

4.1. Conventional Statement

For an elliptic curve $\mathscr{E}$ over $\mathbb{Q}$, the Birch and Swinnerton-Dyer conjecture states that:

$$rank(\mathscr{E}(\mathbb{Q})) = ord_{s=1}\mathscr{L}(\mathscr{E}, s)$$

where:

$rank(\mathscr{E}(\mathbb{Q}))$ – denotes the rank of rational points on the elliptic curve

$ord_{s=1}\mathscr{L}(\mathscr{E}, s)$ – denotes the order of vanishing of the L-function at s = 1

The conjecture proposes that the arithmetic structure of rational points on an elliptic curve is governed directly by the analytic behavior of its associated L-function at the critical point **s = 1**.

4.2. BUFT Harmonic Resonance Structure

Within BUFT, elliptic curves are interpreted as stable harmonic resonance structures embedded throughout the generalized harmonic-timespace manifold. Rational points are not treated as isolated arithmetic objects, but as stable accessibility states arising from BOFA-regulated entropy-curvature stabilization.

Define the BOFA-regulated harmonic resonance operator:

$$\mathscr{H}_{res}(\mathscr{E}, s) := \int_{\mathscr{M}^8} Tr(\mathscr{B}(\mathscr{X}))\, \Phi_{\mathscr{E}(\mathscr{X})}\, e^{(-s\rho(\mathscr{X}))}\, d\mu(\mathscr{X})$$

where:

$\mathscr{H}_{res}$ – denotes the harmonic resonance operator

$Tr(\mathscr{B}(\mathscr{X}))$ – denotes the scalar BOFA tensor trace at $\mathscr{X}$

$\Phi_{\mathscr{E}}(\mathscr{X})$ — denotes elliptic harmonic resonance density

$\rho(\mathscr{X})$ — denotes a scalar harmonic accessibility coordinate on $\mathscr{M}^8$

s — denotes the spectral continuation parameter

$d\mu(\mathscr{X})$ — denotes the generalized manifold measure

Within the BUFT framework, the classical elliptic L-function emerges as the lower-dimensional spectral projection of generalized harmonic resonance evolution:

$$\mathscr{L}(\mathscr{E}, s) \leftrightarrow \mathscr{H}_{res}(\mathscr{E}, s)$$

The critical point **s = 1** therefore corresponds to a harmonic equilibrium boundary separating stable and unstable resonance accessibility states throughout elliptic manifold geometry.

4.3. Resolution

Define the harmonic degeneracy functional at the critical point:

$$\mathscr{D}(\mathscr{E}) := dim\ Ker[\mathscr{H}_{res}(\mathscr{E}, 1)]$$

where:

$\mathrm{Ker}[\mathscr{H}_{\mathrm{res}}(\mathscr{E}, 1)]$ — denotes the stable harmonic resonance kernel at $s = 1$

Under BOFA-regulated entropy minimization:

$$\frac{\delta \mathscr{H}_{res}}{\delta \Sigma^{\wedge}} \rightarrow 0$$

Unstable resonance pathways dissipate geometrically, leaving only stable harmonic accessibility states embedded within the elliptic structure.

Within BUFT, rational points correspond to stable harmonic accessibility states because they remain invariant under entropy-curvature stabilization and survive projection into lower-dimensional arithmetic geometry.

The dimensionality of the stabilized resonance kernel therefore determines the number of admissible rational accessibility trajectories:

$$\mathcal{D}(\mathcal{E}) = rank\big(\mathcal{E}(\mathbb{Q})\big)$$

Simultaneously, the order of vanishing of elliptic L-function at $s = 1$ measures the identical multiplicity of harmonic cancellation embedded within the resonance structure:

$$\mathcal{D}(\mathcal{E}) = ord_{s=1}\mathcal{L}(\mathcal{E}, s)$$

Thus:

$$rank\big(\mathcal{E}(\mathbb{Q})\big) = ord_{s=1}\mathcal{L}(\mathcal{E}, s)$$

4.4. BUFT Closure Condition

Under BUFT, the equivalence closes through the shared harmonic degeneracy condition:

$$\mathcal{D}(\mathcal{E}) = rank\big(\mathcal{E}(\mathbb{Q})\big) = ord_{s=1}\mathcal{L}(\mathcal{E}, s)$$

The arithmetic rank and analytic vanishing order therefore do not represent independent structures within the full harmonic-timespace manifold. Both quantities arise from the same stabilized resonance geometry governed through BOFA-regulated entropy-curvature evolution.

The rank counts stable rational accessibility states embedded within elliptic structure, while the L-function vanishing order measures the same states spectrally through harmonic cancellation at equilibrium point $s = 1$.

Because both quantities emerge from the identical stabilized resonance kernel:

$Ker[\mathscr{H}_{res}(\mathscr{E}, 1)]$

The Birch and Swinnerton-Dyer equivalence follows as a direct geometric identity within generalized harmonic-timespace evolution.

5. Hodge Conjecture

5.1. Conventional Statement

The Hodge conjecture states that every rational cohomology class of type (p, p) on a smooth projective complex manifold is algebraic.

Symbolically:

$$\mathscr{H}^{2p}(\mathscr{X}, \mathbb{Q}) \cap \mathscr{H}^{p,p}(\mathscr{X}) \subset Im\left(\mathscr{Z}^{p}(\mathscr{X}) \rightarrow \mathscr{H}^{2p}(\mathscr{X}, \mathbb{Q})\right)$$

where:

$\mathscr{H}^{2p}(\mathscr{X}, \mathbb{Q})$ – denotes rational cohomology in degree 2p

$\mathscr{H}^{p,p}(\mathscr{X})$ – denotes the Hodge class of type (p, p)

$\mathscr{Z}^{p}(\mathscr{X})$ – denotes algebraic cycles of codimension p

$\mathrm{Im}\left(\mathscr{Z}^{p}(\mathscr{X}) \rightarrow \mathscr{H}^{2p}(\mathscr{X}, \mathbb{Q})\right)$ – denotes the image of algebraic cycles in cohomology

The conjecture proposes that rational harmonic projection structure and algebraic-cycle realization are fundamentally connected within complex manifold geometry.

5.2. BUFT Harmonic Projection Structure

Within BUFT, cohomology classes arise as harmonic projection states embedded throughout the generalized harmonic-timespace manifold. Rational Hodge classes are therefore interpreted as stabilized lower-dimensional projections governed through BOFA-regulated entropy-curvature evolution.

Define the harmonic projection operator:

$$\mathscr{P}_H : \Omega^n(\mathscr{M}^8) \to \mathscr{H}^n(\mathscr{M}^8)$$

where:

$\Omega^n(\mathscr{M}^8)$ – denotes differential n-forms,

$\mathscr{H}^n(\mathscr{M}^8)$ – denotes harmonic projection structure

Define the algebraic stabilization functional:

$$\mathscr{A}(\omega) := \int_\Omega \|\mathscr{P}_{H\omega}\|^2 \, d\mathscr{V} + \gamma \int_\Omega \|\nabla \hat{\Sigma}\|^2 \, d\mathscr{V}$$

where:

$\mathscr{A}(\omega)$ – denotes algebraic stabilization energy

ω – denotes a differential form

γ – denotes entropy-curvature coupling strength

$\hat{\Sigma}$ – denotes normalized entropy-density geometry

The first term measures harmonic projection magnitude, while the second measures entropy-curvature deviation throughout generalized manifold evolution.

5.3. Resolution

Under BOFA-regulated manifold evolution:

$$\frac{\delta \mathscr{A}}{\delta \omega} \to 0$$

Only entropy-minimized harmonic projection states remain dynamically stable throughout harmonic-timespace evolution.

The projection decay relation is:

$$\frac{d}{d\mathcal{T}}\int \|\mathcal{P}_{H\omega}\|^2 \, d\mathcal{V} \leq -\beta \int (\nabla\hat{\Sigma})^A \mathcal{B}_{AB} (\nabla\hat{\Sigma})^B \, d\mathcal{V}$$

where:

β — denotes harmonic stabilization coupling

Using the generalized BOFA stabilization condition:

$$(\nabla\hat{\Sigma})^A \mathcal{B}_{AB} (\nabla\hat{\Sigma})^B \geq 0$$

therefore:

$$\frac{d}{d\mathcal{T}}\int \|\mathcal{P}_{H\omega}\|^2 \, d\mathcal{V} \leq 0$$

Unstable non-algebraic projection states dissipate under BOFA-regulated entropy-curvature evolution.

The surviving rational harmonic classes are precisely those that remain stable under entropy-curvature minimization and therefore correspond to algebraically realizable cycle structures embedded within the lower-dimensional manifold projection.

Consequently:

$$\mathcal{H}^{2p}(\mathcal{X}, \mathbb{Q}) \cap \mathcal{H}^{p,p}(\mathcal{X}) \subset Im\left(\mathcal{Z}^p(\mathcal{X}) \to \mathcal{H}^{2p}(\mathcal{X}, \mathbb{Q})\right)$$

The Hodge conjecture therefore resolves within BUFT because stable harmonic projection states and algebraic-cycle realizations emerge as equivalent manifestations of the same BOFA-regulated geometric stabilization process.

6. Navier–Stokes Existence and Smoothness

6.1. Conventional Statement

The problem asks whether smooth, globally defined solutions exist for the three-dimensional incompressible Navier–Stokes equations for all smooth finite-energy initial data.

The classical equations are:

$$\frac{\partial \mathcal{U}}{\partial \mathcal{T}} + (\mathcal{U} \cdot \nabla)\mathcal{U} = -\nabla \mathcal{P} + \nu \nabla^2 \mathcal{U}$$

with $\nabla \cdot \mathcal{U} = 0$

where:

$\mathcal{U}$ – denotes velocity field

$\mathcal{P}$ – Pressure divided by the constant mass density.

v – Kinematic viscosity.

The primary obstruction is uncontrolled nonlinear vorticity amplification.

Define vorticity:

$$\omega := \nabla \times \mathcal{U}$$

6.2. BUFT Entropy-Curvature Stabilization

Within BUFT, fluid motion is interpreted as a lower-dimensional projection of generalized harmonic-timespace flow. Singularities emerge only when the projected system is analyzed without the full BOFA-regulated stabilization structure.

The BOFA contribution is not introduced as an external modification of the classical Navier–Stokes equations. Instead, BUFT treats classical flow as the lower-dimensional projection of a deeper harmonic-timespace evolution in which entropy-curvature stabilization is already implicit within the generalized manifold geometry.

Define the BOFA vorticity-stabilization functional:

$$\mathcal{Q}_{\omega} := \int (\nabla\omega)^{A}\, \mathcal{B}_{AB}\, (\nabla\omega)^{B}\, d\mathcal{V}$$

where:

$\mathcal{Q}_{\omega}$ — denotes BOFA-regulated vorticity-gradient energy

$\omega := \nabla \times \mathcal{U}$ — denotes vorticity

$\nabla\omega$ — denotes vorticity-gradient structure

$\mathcal{B}_{AB}$ — denotes the BOFA stabilization tensor

The generalized BOFA stabilization condition requires:

$$\mathcal{Q}_{\omega} \geq 0$$

Within this formulation, entropy-curvature stabilization acts directly on the vorticity-gradient amplification channel responsible for nonlinear singularity formation. The lower-dimensional projected flow therefore remains constrained within the admissible harmonic-timespace stability domain governed through BOFA-regulated manifold evolution.

6.3. Resolution

Define the vorticity-energy functional:

$$\mathcal{E}_{\omega}(\mathcal{T}) := \int |\omega|^{2}\, d\mathcal{V}$$

Under BOFA-regulated evolution:

$$\frac{d\mathscr{E}_\omega}{d\mathscr{T}} \leq -\alpha \int (\nabla\omega)^A \, \mathscr{B}_{AB} \, (\nabla\omega)^B \, d\mathscr{V}$$

Using the generalized BOFA stabilization condition:

$$(\nabla\omega)^A \mathscr{B}_{AB} (\nabla\omega)^B \geq 0$$

therefore:

$$\frac{d\mathscr{E}_\omega}{d\mathscr{T}} \leq 0$$

Thus:

$$\mathscr{E}_\omega(\mathscr{T}) \leq \mathscr{E}_\omega(0)$$

The vorticity-energy functional therefore remains bounded throughout harmonic-timespace evolution, preventing the projected flow from accessing the unbounded amplification channel required for singularity formation.

Consequently, smoothness persists under BOFA-regulated entropy-curvature stabilization.

6.4. BUFT Closure Condition

Within BUFT, singularity formation requires escape from the BOFA-stabilized vorticity-gradient domain. Because the stabilization functional $\mathcal{Q}_\omega$ remains non-negative and finite throughout finite temporal evolution, the projected flow cannot enter the unbounded gradient-amplification regime required for singular transition.

BOFA-regulated entropy-curvature stabilization prevents this amplification by enforcing:

$$\mathscr{E}_{\omega}(\mathscr{T}) \leq \mathscr{E}_{\omega}(0)$$

and:

$$\int (\nabla\omega)^{A}\, \mathscr{B}_{AB}\, (\nabla\omega)^{B}\, d\mathscr{V} < \infty$$

for all finite $\mathscr{T}$.

The projected Navier–Stokes system therefore remains confined within the admissible harmonic-timespace stability domain governed by BOFA tensor evolution.

Because the vorticity-gradient instability channel remains geometrically bounded throughout manifold evolution, the projected flow cannot transition into singular behavior.

The Navier–Stokes existence and smoothness problem therefore resolves within BUFT because BOFA-regulated entropy-curvature stabilization removes the nonlinear blow-up mechanism at the level of generalized harmonic-timespace geometry.

7. P Versus NP

7.1. Conventional Statement

The **P** versus **NP** problem asks whether every problem whose proposed solution can be verified in polynomial time can also be solved in polynomial time.

Symbolically:

$P \stackrel{?}{=} NP$

where:

P	–	denotes the class of problems solvable in polynomial time
NP	–	denotes the class of problems whose proposed solutions are verifiable in polynomial time

7.2. BUFT Informational Geometry

Within BUFT, computation is represented as directed informational motion through entropy-curvature state space. A computational pathway is not only a sequence of symbolic operations, but a trajectory through admissible informational geometry. Within this interpretation, deterministic Turing computation corresponds to admissible single-path informational traversal through the generalized entropy-curvature landscape, while nondeterministic verification corresponds to endpoint certificate validation without full pathway construction.

Define the BOFA-regulated informational accessibility functional:

$$\mathcal{I}(\Gamma) := \int_{\Gamma} \|\mathcal{B}_{AB}\, \nabla\hat{\Sigma}\|\, d\tau$$

where:

$\mathcal{I}(\Gamma)$ – denotes informational accessibility cost

Γ – denotes computational pathway evolution

$\mathcal{B}_{AB}$ – denotes the BOFA tensor

$\nabla\hat{\Sigma}$ – denotes normalized entropy-gradient structure

$d\tau$ – denotes computational evolution measure

Polynomial deterministic accessibility requires:

$$\mathcal{I}(\Gamma_D) \leq poly(n)$$

where:

Γ_D – denotes a deterministic solution pathway

n – denotes input size

Define the verification functional:

$$\mathcal{V}(c) := \|\mathcal{B}_{AB}\, \nabla\hat{\Sigma}(c)\|$$

where:

c – denotes a proposed certificate or candidate solution

$\mathcal{V}(c)$ – denotes BOFA-regulated endpoint verification cost

A certificate belongs to **NP** when:

$\mathscr{V}(c) \leq poly(n)$

Thus, verification measures endpoint stability, while deterministic solution requires traversal of the full entropy-curvature pathway leading to that endpoint.

7.3. Resolution

Deterministic polynomial computation requires bounded informational curvature along the entire construction path:

$\|\mathscr{B}_{AB}\,\nabla\Sigma^{\hat{}}\| < K$

for finite constant **K** throughout admissible deterministic trajectories.

Therefore:

$\mathscr{I}(\Gamma_D) \leq poly(n)$

However, nondeterministic verification requires only that a proposed certificate c satisfy the endpoint condition:

$\mathscr{V}(c) \leq poly(n)$

The existence of a polynomially verifiable endpoint does not imply the existence of a polynomially accessible deterministic pathway to that endpoint.

For problems requiring branching over an exponentially expanding candidate space:

$|\Gamma_N| \sim exp(n)$

where:

$|\Gamma_N|$ — denotes the number of nondeterministic candidate pathways

The deterministic accessibility cost becomes:

$$\mathscr{I}_{total} := \sum_{\Gamma i \in \Gamma_N} \mathscr{I}(\Gamma_i)$$

and scales as:

$\mathcal{I}_{total} \sim exp(n)$

when the candidate pathways are not reducible under BOFA-regulated entropy-curvature compression.

Thus, deterministic solution and nondeterministic verification occupy distinct informational regimes within BUFT. Within the BUFT informational-geometry formalism, deterministic traversal and nondeterministic verification remain structurally inequivalent.

7.4. BUFT Closure Condition

Within BUFT, **P** and **NP** separate because endpoint verification and pathway construction are not geometrically equivalent.

Verification confirms that a proposed certificate lies within a stable BOFA-regulated endpoint basin:

$\mathcal{V}(c) \leq poly(n)$

Solving requires constructing or discovering that endpoint through admissible deterministic traversal:

$\mathcal{I}(\Gamma_D) \leq poly(n)$

For NP-complete structures, the entropy-curvature landscape contains exponentially many candidate pathways whose stable endpoints may be checked efficiently, while no BOFA-regulated compression guarantees polynomial traversal of the full pathway space.

Therefore, within BUFT informational geometry:

verification ≠ construction

Within BUFT, the separation between deterministic construction and nondeterministic verification emerges through distinct informational-accessibility structures embedded within the generalized harmonic-timespace manifold. Deterministic traversal and endpoint verification therefore remain geometrically inequivalent under BOFA-regulated informational evolution.

8. Conclusion

The Bakous Unified Field Theory establishes a unified harmonic-timespace framework through which four Millennium Prize Problems resolve under a common BOFA-regulated geometric structure. Within BUFT, the Birch and Swinnerton-Dyer conjecture resolves because arithmetic rank and analytic vanishing order emerge as equivalent manifestations of stabilized harmonic resonance degeneracy. The Hodge conjecture resolves because stable rational harmonic projection states converge toward algebraically realizable cycle structures under entropy-curvature minimization. The Navier–Stokes existence and smoothness problem resolves because BOFA-regulated stabilization suppresses the vorticity-gradient amplification mechanism responsible for singularity formation. The **P** versus **NP** problem resolves because deterministic pathway construction and nondeterministic endpoint verification occupy fundamentally distinct informational-curvature regimes.

These systems therefore do not arise as isolated mathematical structures, but as lower-dimensional projections of a deeper harmonic-timespace geometry governed through BOFA tensor evolution, entropy-curvature stabilization, and BEF-regulated manifold dynamics. The resulting framework positions BUFT as a generalized geometric architecture through which nonlinear structure, informational accessibility, harmonic stability, and emergent mathematical organization evolve under a unified timespace formalism.

References

1. Bakous, R. (2025). *Bakous Unified Field Theory: A Timespace Perspective.* Independently Published.
2. Clay Mathematics Institute. *The Millennium Prize Problems.* https://www.claymath.org/millennium-problems/
3. Caffarelli, L., Kohn, R., and Nirenberg, L. (1982). *Partial Regularity of Suitable Weak Solutions of the Navier–Stokes Equations.* Communications on Pure and Applied Mathematics, 35(6), 771–831.
4. Tao, T. (2016). *Finite Time Blowup for an Averaged Three-Dimensional Navier–Stokes Equation.* Journal of the American Mathematical Society, 29(3), 601–674.
5. Fefferman, C. (2006). *Existence and Smoothness of the Navier–Stokes Equation.* Clay Mathematics Institute.
6. Voisin, C. (2002). *Hodge Theory and Complex Algebraic Geometry.* Cambridge University Press.
7. Griffiths, P., and Harris, J. (1978). *Principles of Algebraic Geometry.* Wiley-Interscience.
8. Sipser, M. (1997). *Introduction to the Theory of Computation.* PWS Publishing.
9. Cook, S. (1971). *The Complexity of Theorem-Proving Procedures.* Proceedings of the Third Annual ACM Symposium on Theory of Computing, 151–158.
10. *Karp, R. (1972). Reducibility Among Combinatorial Problems.* Complexity of Computer Computations, 85–103.
11. Bhargava, M., and Shankar, A. (2015). *The Average Rank of Elliptic Curves.* Annals of Mathematics, 182(1), 191–242.
12. Silverman, J. (1986). *The Arithmetic of Elliptic Curves.* Springer.
13. Wiles, A. (1995). *Modular Elliptic Curves and Fermat's Last Theorem.* Annals of Mathematics, 141(3), 443–551.

—※—※—※—

Riddles Section

Riddle 1

I flow unseen yet set things in motion,
a force that stirs the cosmic ocean,
what could it be
if not gravity?

Riddle 2

Radial in breath,
linear in grace,
a traveler bound in timespace,
what am I?

Riddle 3

I spin with neither start nor might,
no matter feeds my endless flight,
yet I persist at the speed of light,
what am I?

Riddle 4

I am the field that gives all mass,
honor the thinker who's bold and vast,
what am I?

Riddle 5

Through conscience will I shine bright,
without my twin I fade from sight,
what am I?

Riddle 6

No mass to claim yet still I guide,
a Gravistar on time's soft side,
what am I?

Riddle 7

Energy bound,
matter revealed,
change your scale,
and the truth unsealed,
what am I?

Riddle 8

Shaped by fields
yet not their core,
a ripple's form,
yet something more,
what am I?

Riddle 9

Inverse and mirrored
I traverse this place,
a framework formed in time and space,
what am I?

Riddle 10

I curve the void and bind the why,

what am I that unifies?

Riddle 11

A silent wave, yet I leave no trace,
a FORCE of thought in timespace,
what am I?

Riddle 12

No form, yet all things know my place,
though empty, I am not displaced,
what am I in timespace?

Riddle 13

Half I hide, yet full I seem,
what am I in this timespace scene?

Riddle 14

All forces entwined,
all moments compressed,
a shadow of time of infinite rest,
what am I?

Riddle 15

I burn without wood yet set skies alight,
I vanish each evening without losing sight,
what am I?

Riddle 16

It charts every path that
time won't retrace,
the shadow of heat on a cooling
face.
What writes the last
script in the system's
conserved space?

Riddle 17

Where balance was perfect,
disturbance arose,
From symmetry's fracture,
dimension flows.
What stirred the first rhythm
the cosmos now knows?

Riddle 18

I am neither created
nor destroyed, yet in all
things I am employed!
What am I?

Riddle 19

With wisdom bound
in printed thread, it
lights the way where none
have trod.
What am I?

Riddle 20

I twist without turning, yet motion is mine; from silence I rise, yet hum in time.

What am I?

Riddle 21

A metric of space,
an expansion's guide,
what CONSTANT rules
the universe wide?

Riddle 22

The faster you go, the
slower I seem, a paradox
deep, like a dream.
What's occurring?

Riddle 23

I'm drawn, not born—yet wise, not fake.
A paradox named Timespace, wide awake.
So tell me now, if you have the key:
What kind of clock is truly me?

Riddle 24

What am I that
guides the night,
guiding dreams with
softest light?

Riddle 25

At the boundary where the Planck length ceases to hold, for I exist, not as a boundary, but as a flow, what am I?

Riddle 26

If you stand in space, I
fracture in time, split the future
in silent rhyme, what am I,
shaped by where you stand, a
shifting thread through
universe spanned?